The Fertile Fjord

The Fertile Fjord

Plankton in Puget Sound

Richard M. Strickland

With a Foreword by Joel Hedgpeth

A Washington Sea Grant Publication
Distributed by the University of Washington Press
Seattle and London

Cover Photos: The Marine jellyfish *Polyorchis* by
Langdon B. Quetin and Robin M. Ross, Marine Science Institute,
University of California Santa Barbara. Author photo by Dawn J. Prince.

First published in 1983 by
Washington Sea Grant Program
University of Washington

Distributed by University of Washington Press
Seattle, Washington 98195

The following are reprinted by permission of the copyright holders who retain all rights: Plankton © Joan Swift (reprinted from *Poetry Northwest*); photographs © Langdon B. Quetin, Puget Sound Maritime Historical Society, Smyth Associates, Louisa Nishitani, Bruce W. Frost, Brett Dumbauld, Beatrice C. Booth, Susan B. Stanton, Mark D. Ohman, Alexander J. Chester, Charles H. Greene, Richard Kaiser, T. S. English, William D. Waddington, the Friday Harbor Laboratories, and the National Marine Fisheries Service.

Publication of this book was supported by grants (04-5-158-48; 04-7-158-44021; NA79AA-D-00054 and NA81AA-D-00030) from the National Oceanic and Atmospheric Administration and by funds from the Environmental Protection Agency. Writing and publication was conducted by the Washington Sea Grant Program under project A/PC-7.

The U.S. Government is authorized to produce and distribute reprints for governmental purposes notwithstanding any copyright notation that may appear hereon.

Library of Congress Cataloging in Publication Data

Strickland, Richard M., 1950–
   The fertile fjord.

   (Puget Sound books)
   Bibliography: p.
   Includes index.
   1. Marine plankton—Washington (State)—Puget Sound.
I. Title. II. Series: Puget Sound books.
QH91.8.P5S77   1983   574.92'632   82-19995
ISBN 0-295-95979-7

Contents

To my parents, Dick and Margo.
Without their unwavering support,
this book would not have been possible.

About the Puget Sound Books

This book is one of a series of books that have been commissioned to provide readers with useful information about Puget Sound . . .

About its physical properties—the shape and form of the Sound, the physical and chemical nature of its waters, and the interaction of these waters with the surrounding shorelines.

About the biological aspects of the Sound—the plankton that form the basis of its food chains; the fishes that swim in this inland sea; the region's marine birds and mammals; and the habitats that nourish and protect its wildlife.

About man's uses of the Sound—his harvest of finfish, shellfish, and even seaweed; the transport of people and goods on these crowded waters; and the pursuit of recreation and esthetic fulfillment in this marine setting.

About man and his relationships to this region—the characteristics of the populations which surround Puget Sound; the governance of man's activities and the management of the region's natural resources; and finally, the historical uses of this magnificent resource—Puget Sound.

To produce these books has required more than six years and the dedicated efforts of more than one hundred people. This series was initiated in 1977 through a survey of several hundred potential readers with diverse and wide-ranging interests.

The collective preferences of these individuals became the standards against which the project staff and the editorial board determined the scope of each volume and decided upon the style and kind of presentation appropriate for the series.

In the Spring of 1978, a prospectus outlining these criteria and inviting expressions of interest in writing any one of the volumes was distributed to individuals, institutions, and organizations throughout Western Washington. The responses were gratifying. For each volume no fewer than two and as many as eight outlines were submitted for consideration by the staff and the editorial board. The authors who were subsequently chosen were selected not only for their expertise in a particular field but also for their ability to convey information in the manner requested.

Nevertheless, each book has a distinct flavor—the result of each author's style and demands of the subject being written about. Although each volume is part of a series, there has been little desire on the part of the staff to eliminate the individuality of each volume. Indeed, creative yet responsible expression has been encouraged.

This series would not have been undertaken without the substantial support of the Puget Sound Marine EcoSystems Analysis(MESA) Project within the Office of Oceanography and Marine Services/Ocean Assessment Division of the National Oceanic and Atmospheric Administration. From the start, the representatives of this office have supported the conceptual design of this series, the writing, and the production. Financial support for the project was also received from the Environmental Protection Agency and from the Washington Sea Grant Program. All these agencies have supported the series as part of their continuing efforts to provide information that is useful in assessing existing and potential environmental problems, stresses, and uses of Puget Sound.

Any major undertaking such as this series requires the efforts of a great many people. Only the names of those most closely associated with the Puget Sound Books—the writers, the editors, the illustrators and cartographers, the editorial board, the project's administrators and its sponsors—have been listed here. All these people—and many more—have contributed to this series, which is dedicated to the people who live, work, and play on and beside Puget Sound.

Alyn Duxbury and Patricia Peyton
June 1983

Foreword

In the 1950s, during the first flowering of oceanography after World War II, great public interest and expectation was aroused in the Sunday supplements and more substantial magazines for a future of limitless food from the sea. The vast hordes of plankton were an untapped resource, and certain plankton organisms, especially *Chlorella* would be cultured with ease and economy. As Lionel A. Walford remarked, what was expected "was an entirely new kind of food that can be produced in massive quantities at neglible cost, which somebody will eat, presumably the backward peoples." One Japanese gentleman, lecturing at Scripps Institution of Oceanography, told us that the product of mass culture of *Chlorella* tasted like soy sauce, "but we Japanese like soy sauce."

It was during this excitement that Carl Sauer granted me permission to audit his course in natural resources (Geography 153) provided I took over his pulpit for a week when he was off to Washington. I wish I had had a book like this about the plankton of Puget Sound from which to crib the material and to place on the class reading list. I had only three days to explain the resources of the sea to a group of students who were only beginning to learn that geography, as Carl Sauer saw it, was not an armchair review of the less substantial articles in National Geographic magazine but a solid part of the science of the environment. After my first lecture on the phytoplankton base of the resources of the sea, one of my friends overheard one of the students in the hallway afterwards remarking "I never heard of this business of little one-celled plants without roots. Does this guy know what he's talking about?"

Things have changed since then, even at Berkeley. We do know enough, I think, to realize that plankton is part of a system, not a harvestable end in itself. The world of the plankton and its intricate internal relationships is not a subject for a brief overview if we are concerned with man's present and future reliance upon the resources of the sea. The National Sea Grant College Program is a response to the need for understanding the resources of the sea and how to utilize them in the future. This book on the plankton of Puget Sound meets that need admirably, and, along with the handbook of marine birds and mammals and future volumes on biology and oceanography (in conjunction with those on history and economics of Puget Sound) will be a most significant contribution of the Washington Sea Grant Program to the people of the region it serves. It should be the envy of, and set the example for, other Sea Grant programs in the nation. Knowledge acquired and stored away in data banks or presented in scattered brochures is not really knowledge. Like the tree that falls in distant Siberia, it makes no sound when there is no one to hear it.

Joel W. Hedgpeth

Preface

Great literature, until the last century or so, glorified only great adventurers and nobility. It ignored the masses of common people that make society function. So it is still with popular studies of the sea—the noble whales and sharks get more attention than the multitudes of tiny organisms that they simultaneously rule and depend upon. This book attempts to redress that imbalance, at least for Puget Sound. The Sound is a bountiful body of water due to a unique coincidence of physical and biological properties, at the center of which is the invisible, anonymous proletariat of the plankton.

This book is intended as a capsule review for scientists and nonscientists with an interest in natural history of plankton and its relationship to the pelagic food web in Puget Sound. It is not intended as a field guide to plankton, several of which already exist; instead, it is a synthesis of ecological knowledge that bridges some conceptual gaps in existing data. Paradoxically, although there is much more information about plankton than reasonably could be presented here, there are also disturbing voids in our knowledge that can be filled only by further research. If readers are left with unanswered questions and an appetite for more information, the book will have achieved its goal of capturing interest in this fascinating but often overlooked subject.

Richard Strickland
June 1983

Acknowledgments

This book represents the enormous efforts of many people besides myself, especially Washington Sea Grant staff members Andrea Jarvela, Cathy Lauenstein, Patricia Peyton, Alyn Duxbury, Kirk Johnson, Diana Jensen, and Laura Mason. Valuable contributions came from the careful reviews of Karl Banse, Mark Ohman, Louisa Nishitani, and Eugene Collias of the University of Washington, Kurt Fresh of the Washington Department of Fisheries, and Puget Sound Books Editorial Board members Michael Waldichuk and Elizabeth Twiss. George C. Anderson of the University of Washington helped obtain the contract and offered encouragement at the outset, as did Alexander J. Chester (now of Beaufort, North Carolina) and Kathleen J. Smith, M.D., (now of Durham, Pennsylvania). The title was graciously provided by Cynthia Marks of the University of Washington.

Many people generously discussed the text and helped develop illustrations: Kenneth Adkins, Keith Benson, Kendra Daly, T. Saunders English, Noel McGary, Kathy Newell, Mark Ohman, James Postel, Jeanette Yen, and the staff of the Fisheries-Oceanography Library of the University of Washington; Kevin Bailey, Edward Baker, Robert Burns, David Damkaer, Richard Feely, Howard Harris, and Marilyn Lamb of the National Oceanic and Atmospheric Administration; Dale Anderson, Michael Bertman, Robert Dexter, and Elizabeth Quinlan with whom I worked at the URS Company, Seattle; Steve and Jan Smyth of Smyth Associates Inc., Kirkland; and Buzz Shaw of the Seattle Aquarium. Some electromicrographs appear courtesy of Northwest and Alaska Fisheries Center, National Marine Fisheries Service, NOAA, Seattle. Some unpublished data appear courtesy of Municipality of Metropolitan Seattle (METRO).

A special thanks to Denise Krouse and to all my friends in Seattle for their moral support.

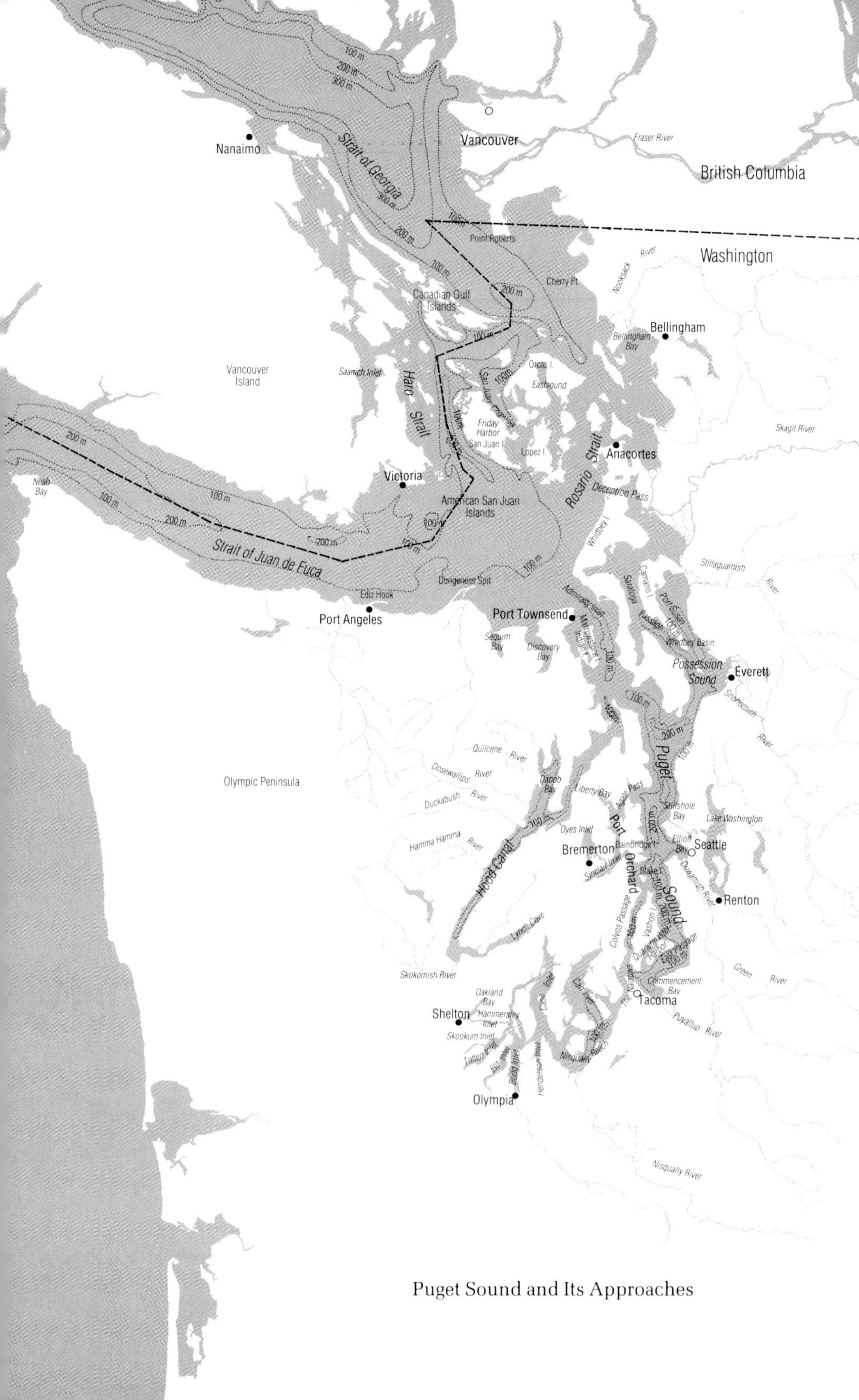

Puget Sound and Its Approaches

Plankton

Sometimes, at night, each one
has a lightning bug in it.
That's how you see them,
invisible
under and around the boat.
I drop a pail over the side

think how for herring
they are three meals,
phosphorescence
in the flesh of bass
and the deep blueness of whales.

They live their lives unseen,
not just gray mobs without faces
but like calm steady workers
in some underground plot
to keep the world alive.

I stare into the pail
where thousands drift.
When the sky is dark enough
I'll row the dinghy out and lift
oarsful of their dripping light.
I won't miss the sun on the other side
because it is here,
brushed angel wings
when I dive and float
flapping my arms.

Later, I'll towel them from my skin,
taste salmon,
oysters,
some of the little light.

Joan Swift

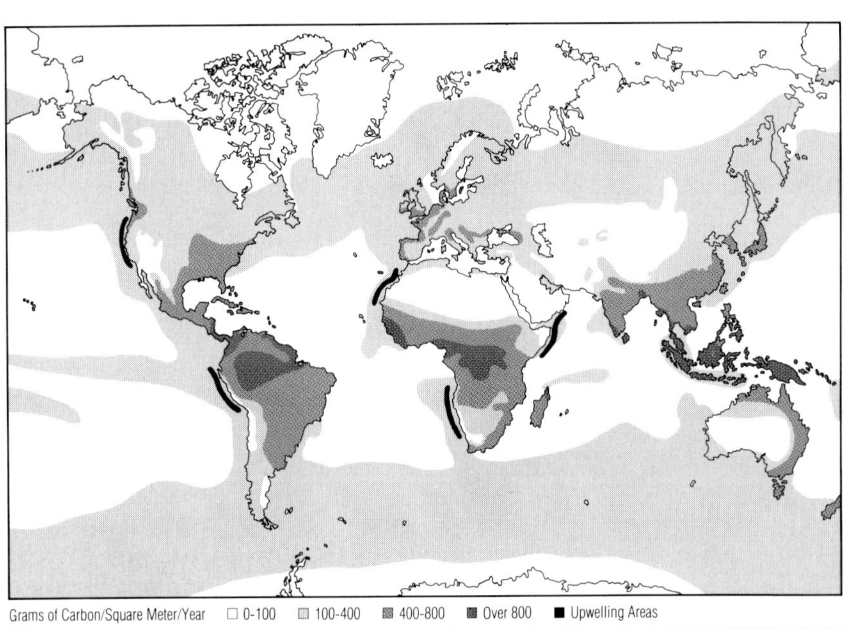

Figure 1.1 Global distribution of plant productivity, based on average annual production of plant matter per unit area. The most productive regions are marshes, reefs, and tropical forests. The open oceans are as unproductive as terrestrial deserts. Coastal waters, on the other hand, are where planktonic plants are most productive and support 99 percent of the world's commercial fish catch. Half of that catch comes from areas where plankton growth is stimulated by upwelling of water, as also occurs in Puget Sound. (After Rand McNally, 1977)

CHAPTER ONE

Plankton Primer

The role of the infinitely small in nature is infinitely great.
Louis Pasteur

The most important animals in Puget Sound, as in the ocean at large, are neither the most obvious nor the most beloved. They are not the whales, nor the salmon, nor the clams, crabs, or octopi. Neither are seaweeds the most important plants in the Sound. Floating freely with the tides and currents, little known and nearly invisible, are the tiny plants and animals of the plankton. What these organisms lack in size and glamour, they compensate for in importance to all marine life.

The larger, better known, and seemingly more important marine organisms are barely significant to the living yield of the sea as a whole, and their absence, though regrettable, would not threaten the grander web of life on earth. Steinbeck said:

> The disappearance of plankton, although the components are microscopic, would probably in a short time eliminate every living thing in the sea and change the whole of man's life, if it did not through a seismic disturbance of balance eliminate all life on the globe.

We have since learned that terrestrial life could continue without the plankton, and that some exotic animals subsist only on hot chemicals belching from the sea bottom, independent of other marine life. Yet despite these discoveries, plankton still is to the ocean what grasses and flowers, insects and rodents are to the land. Fishes fatten on swarms of diminutive animals grazing in pastures of microalgae, all suspended in a four-dimensional fluid world. Indeed, a diet of planktonic krill sustains the largest creature ever to live, the blue whale.

Plankton dominates the biological budget of the sea. Of all the marine plant and animal tissue produced beyond the immediate fringes of the shoreline and the bottom, more than 90 percent is plankton. This is particularly true in Puget Sound, where the rate of plant production ranks among the highest in saltwater environments, rivaling that of terrestrial forests and farms (Figure 1.1 and 1.2). Furthermore, the yield to humans of fish and shellfish is governed in large part by the growth of the plankton on which these animals or their prey depend. The oceans cover two-thirds of the planet, but virtually all commercial fish tonnage comes from the relatively shallow areas flanking the continents, where plankton growth is most vigorous.

To the biologist who looks at the entire marine world, plankton is clearly the principal life form, and Puget Sound is an ideal place to study it. This book will describe the unique characteristics of plankton that make it abundant, ubiquitous, and vital to life in Puget Sound. Far from being exotic and abstruse, plankton research is the very heart of marine biology.

Plankton—The Drifters

"Plankton" is derived from a Greek word which means "free-floating." While many plankters can swim as well as float, plankton includes any aquatic—in this case, marine—organism living unattached and having swimming power insufficient to resist most water currents. But there is more to the definition, as described by Hardy:

> "Plankton" is one of the most expressive technical terms used in science and is taken directly from the Greek. It is often translated as if it meant just "wandering," but really the Greek is more subtle than this and tells us in one word what we in English have to say in several; it has a distinctly passive sense meaning "that which is made to wander or drift," i.e. drifting beyond its own control—unable to stop if it wanted to.

Plankton includes bacteria, plants, animals, and even some organisms that seem both plant and animal. In keeping with their weak mobility, most planktonic organisms are quite small; few are larger than a common housefly.

The term phytoplankton refers to those members of the plankton which make their own food from sunlight. All phytoplankters (except perhaps some photosynthetic bacteria uncommon in Puget Sound) belong to that group of plants called algae, which also includes seaweeds. Non-photosynthetic bacteria that live suspended in the water are called bacterioplankton. All other plankters—that is, all of the planktonic animals—are referred to as zooplankton.

Plankton contains an extremely diverse range of organisms; nearly every major group of animals has representatives in the zooplankton. Many creatures exist in the plankton only until they grow large enough to swim independently (such adults are called nekton), or until they are transformed to live as adults (the benthos) on the sea bottom. These temporary zooplankters—larval stages of such animals as crabs, starfishes, clams, and some fishes—are called meroplankton. Organisms remaining in the plankton their entire lives are called holoplankton.

Why Plankton Is Important

"Plankton" is an unusual term in biology—it classifies organisms according to their locomotion rather than their genetic kinships. The unique constraints and consequences of the planktonic lifestyle merit separate consideration in science. The most significant roles of the

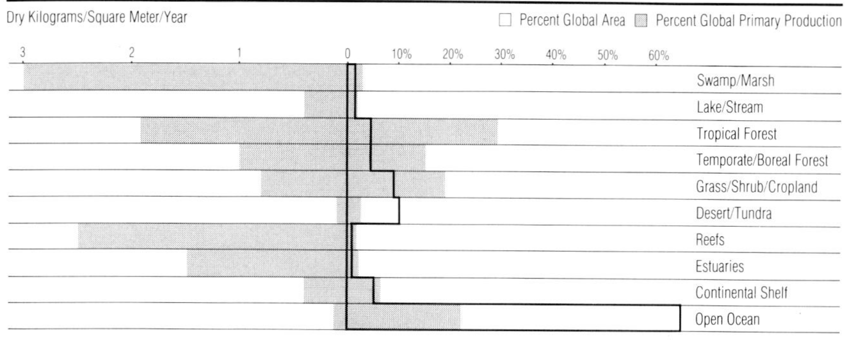

Figure 1.2 Average annual production of plant matter in different eco-systems shown by two measures: dry yield per unit area and the fraction each contributes to worldwide yield based on its global area. The oceans are mostly unproductive except near the coasts, hence the two-thirds of the earth covered by water contributes only one-third of global plant production. (After Whittaker, 1975).

plankton are as sources of food, oxygen, and energy, as ecological and paleoecological indicators, and as poison.

Food Source

Plankton is the nutritional foundation of the sea's biological edifice. The breadbasket of the sea is the phytoplankton; phytoplankton is food for zooplankton, and zooplankton is food for fishes, leading finally to birds, mammals, and humans. Furthermore, the corpses and non-living organic by-products of all these organisms sink and sustain animals on the sea bottom. Outside of some in the shallowest few meters of water close to shore, virtually all marine animals are ultimately dependent on plankton. Some people believe that plankton can become a food for humans as well.

Oxygen Source

All of the oxygen that animals and plants respire is generated by the photosynthesis of plants, from the splitting of water molecules using solar energy. Photosynthesis also removes excess carbon dioxide from the atmosphere. Primitive phytoplankton in the primordial sea changed the world forever by introducing oxygen into the atmosphere over a billion years ago. Phytoplankton today produces 95 percent of the oxygen generated in the sea, and a third of the earth's oxygen (oxygen of terrestrial origin could still sustain the earth if plankton disappeared). While most marine oxygen comes from the expanses of the open oceans, areas such as Puget Sound are more than ten times as productive for their area (Figures 1.1 and 1.2).

It is also a paradox that these organisms that supply oxygen sometimes also cause oxygen shortages in waters of the Puget Sound area. The decay of dead phytoplankton can strip the oxygen from stagnant waters. In a few places in Puget Sound, following rapid outbursts of phytoplankton growth probably aggravated by pollution and other human activities, acute episodes of oxygen depletion cause problems for fishes and other animals. Some inlets in the area also have natural, chronic oxygen shortages to which certain animals seem to have adapted.

Energy Source

Settled to the ocean bottom, buried, compressed, and heated under just the right geological conditions for millions of years, plankton becomes the viscid stew we now tap as petroleum. In crude oil are found hydrocarbons, mostly derived from the lipids of phytoplankton, that are the remnants of blooms millenia ago. Traces of plant pigments still appear in oil. Fossilized zooplankton droppings are found in oil shale deposits, and many of the components of oil have properties, such as the predominance of light isotopes of carbon and the ability to rotate the plane of polarized light, that can be directly related to planktonic origins.

Oil today is generated continuously on the ocean floor by the liquefaction of plankton sediments, but this renewable resource is far from boundless. If all the plankton produced in the world in a year became oil (which it does not), it would supply the United States—assuming a rate of consumption of nearly 20 million barrels a day—for less than a week. Oil is found where plankton was most productive during geologic history, along the ancient continental shelves, reefs, and inland seas and lakes. Puget Sound might be an outstanding location for drilling operations in a few million years, if it isn't wiped out by another ice age.

Ecological and Paleoecological Indicator

Geologists obtain from fossils information about the nature of the environments in which fossils form and the relative ages of the rocks containing them. Certain planktonic species are known to have existed only at certain geologic times, and to have been associated with known sets of conditions. By analyzing the shapes and chemical makeup of tiny shells left by plankters millions of years ago, paleontologists can trace ancient fluctuations in global climate, and chart historical changes in ocean currents and salinity.

Biologists can also tell a great deal about today's water bodies from their plankton communities. A particular composition of plankton is a signal, for example, of deterioration of water quality. The eutrophication of Lake Washington due to sewage during the 1960s was diagnosed by the appearance of eutrophic-type plankters in time to avert widespread beach closures and loss of fish populations. Lake Washington is now being rehabilitated, as demonstrated by the return of plankton that is characteristic of cold, clear, clean lakes.

Poison

In the autumn of 1978 several persons became ill after eating shellfish gathered from Whidbey Island beaches. State health officials sampled the shellfish and immediately closed the affected beaches, because of the presence of paralytic shellfish poisoning (PSP). The illness is caused by toxins formed by a species of phytoplankton and accumulated in the tissues of mussels, clams, and oysters feeding on the plankton. Such outbreaks are called "red tides." Plankton sometimes does grow densely enough to tint the water red, but the relationship between red tides and PSP is not so simple.

PSP toxins are some of the many chemicals—including ordinary mineral nutrients—that can be transferred from plankton to higher animals. Of particular concern are those toxins which concentrate in living tissues. Now banned in the United States, the insecticide DDT is a classic example. DDT was found in higher concentrations in fishes than in the plankton they eat, and in some locations reached harmful levels in certain fish-eating birds. The plankton occupies a crucial role in the pathways by which the growing number of such industrial chemicals now finding their way into the marine environment reach higher animals and people.

The chapters that follow will elaborate on the roles of plankton in food production and in pollution in Puget Sound. First, however, we will gain some familiarity with plankton, the world it inhabits, and how scientists study it.

CHAPTER TWO

Studying Plankton

The first requirement of a scientist is that he be curious. He should be capable
of being astonished and eager to find out.

Erwin Schroedinger

The waxing and waning of plankton in Puget Sound are scarcely visible to the naked eye. We cannot see that plankton swarms drift across its surface like clouds across the sky. At times we may see faint gradations of color, from wintry blue to muddy brown to spring green, or occasional prominent red tides, but most of the affairs of the plankton are as private as those of deep-sea creatures. SCUBA divers have a closer perspective, but what they gain from proximity, they lose in breadth and depth of vision. Fortunately, with the aid of ships and instruments, we can crudely construct what our eyes fail to perceive directly.

Sampling

Sampling is the central art of the outdoor sciences, and of biological oceanography in particular. What is known about plankton is dictated ultimately by the act of sampling, for we can learn little about organisms we cannot capture. A plankton sample is a window for viewing microscopic sea life. If some plankters escape sampling, then that view is obscured or distorted.

Plankton nets are used to capture and concentrate organisms living in a dilute suspension in the upper layers of sea. Used by fishermen for centuries, nets were adapted to fish for microorganisms only with the advent of the microscope. In 1828 J. Vaughn Thompson of Cork, Ireland sewed silk cloth of the type used for sifting flour into a cone. When pulled through the water, the cloth strained out tiny organisms and deposited them in an open jar sewn onto the pointed end. Thompson called his device a tow-net, and modern plankton nets are hardly different. Charles Darwin used a tow-net aboard the *Beagle*, on the cruise which stimulated his 1859 *Origin of Species*. Tow-nets were also standard equipment aboard the *H.M.S. Challenger*, which departed England in 1872 for four years at sea, on a voyage commonly regarded as marking the birth of modern oceanography.

To Darwin and his peers, nets and microscopes revealed nothing more than smaller members of the known groups of plants and animals. It was not until 1887 that Victor Hensen of Kiel, Germany thought of the organisms caught in tow-nets as a distinct biological community and coined the term plankton. It was an innovative way to classify organisms—by their habitat and locomotion rather than by their ancestry—and it grew out of sampling technology.

No single net can faithfully sample all types of plankton. Plankton nets come in two varieties, those used for capturing phytoplankton and zooplankton. Phytoplankton nets, now seldom used, are small with a fine mesh for capturing abundant (up to several million per liter) tiny cells. Zooplankton nets are larger and coarser-meshed, for pursuing larger and scarcer organisms (Figure 2.1). They offer less resistance when pulled through the water, and the mesh is large enough to allow phytoplankters and some of the smaller animals to pass through to reduce clogging. This permits the net to be towed at the higher speeds necessary to capture large animals with some swimming ability.

Numbers and types of plankters differ from surface to deep water. The earliest plankton nets could not realistically sample deep-water plankton, because the open net would be contaminated with surface

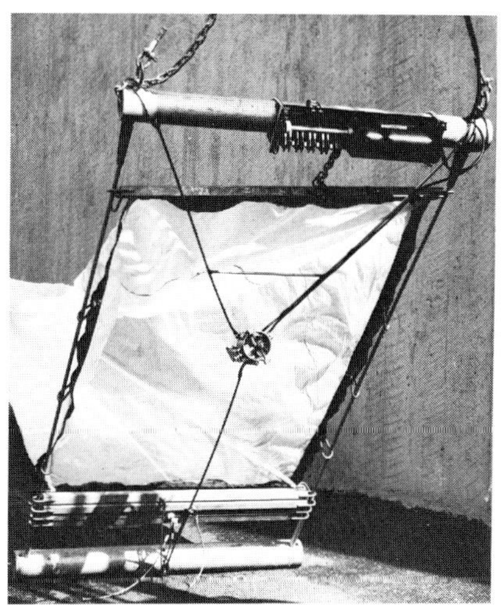

Figure 2.1 The simplest and oldest zooplankton net (below) is a cone about one meter across, six meters long, with 0.3–0.6 millimeter mesh. The modern Tucker trawl (left) is two meters square, has a flow meter, and is really several accordion-like nets opened by electronic signals to sample sequential depth strata. (Courtesy B. W. Frost)

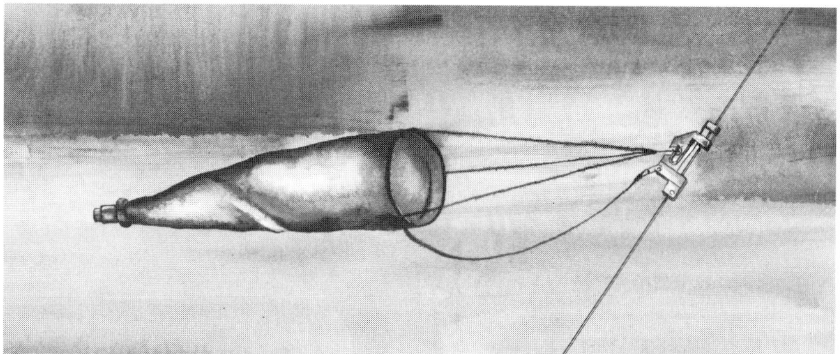

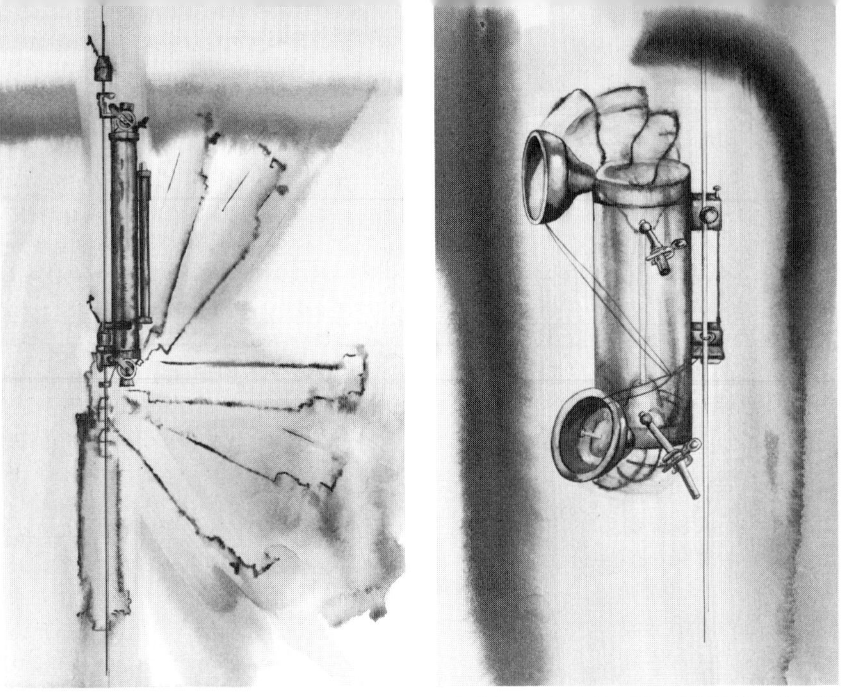

Figure 2.2 An early standard sampling bottle (left) was named for Norwegian explorer Fridtjof Nansen, who froze his ship in arctic ice for three years to follow the currents. The one-liter Nansen bottle, made of galvanized steel, inverts to close and trap the sample. The Van Dorn bottle (right) is larger and made of nontoxic plastic to reduce inhibition of plankton growth. It is closed by a weight that releases two rubberband-loaded plugs.

plankton as it was retrieved. Thus an early improvement to net design was a capability for closing the net and isolating deep-water samples.

A modern net may also have a flow meter to measure the volume of water sampled, permitting a count of animals per unit volume. Small zooplankton nets can capture animals the size of copepods, while larger and faster midwater trawls must be used to capture the strongest swimmers, such as planktonic shrimps and larval fishes.

Two recent innovations have brought the zooplankton net to a high state of sophistication. The Tucker trawl has several compartments which can be opened and closed automatically at different depths by an electronic signal. It can thus provide sequential pictures of the zooplankton communities in several depth strata. With any net, however, it is difficult to determine whether some animals, perhaps warned by the vibrations of the cable towing the net, are escaping capture. A new net developed on Puget Sound avoids this problem by sampling while it sinks under its own weight, instead of while being towed surfaceward.

For sampling phytoplankton, nets have been largely replaced by closable bottles (Figure 2.2). The advantage of using bottles is that they capture even the tiniest cells, which slip through nets. Furthermore, sample depth and volume can be accurately determined, several depths may be sampled simultaneously, and water is collected for chemical analysis along with the plankton. The disadvantage is that bottle samples do not concentrate the plankton as nets do. This must be done after collection, most commonly by killing the plankters with formaldehyde or iodine in a special chamber, and examining them after they sink onto a microscope slide that forms the chamber bottom.

A third method of sampling is to pump water directly aboard ship. In many ways this is an ideal method—a hose is lowered to the desired depth, the pump is switched on, and liter after liter of water is instantly available. Pumps and hoses are difficult to handle in rough seas or at great depths, however, and even the best pump processes only a fraction of the volume sampled by a large zooplankton net. Modern vessels use pumps mainly as a convenient method of surveying a large area quickly.

The most serious problem with plankton sampling is the artificiality of observing plankton out of its natural element. In describing the first observations of deep-sea creatures—including plankton—from a submersible vehicle, and reflecting on the shackles which bind most marine scientists to the surface, Jacques Cousteau said: "Oceanographers, bless them, are blind beggars, tottering about on crutches of cables." Indeed, some of the most dramatic recent discoveries about plankton—the habits of delicate gelatinous zooplankters, for instance—have resulted when oceanographers immersed themselves in the sea, and observed directly what their crude sampling devices had for decades been destroying in the process of capture. This problem is common to all sciences; the process of observation alters what the scientist wants to observe.

Early Research on Puget Sound

The first published scientific observations of plankton on Puget Sound were conducted on jellyfish (Figure 2.3) by Louis Agassiz and his son Alexander, who cruised the area in the summer of 1859. The elder Agassiz, America's foremost naturalist, also founded Harvard's Museum of Comparative Zoology that year, and in 1873 set up America's first marine lab on Penikese Island off Cape Cod.

After his father's death, Alexander Agassiz returned to Puget Sound several times between 1888 and 1905, aboard the renowned U.S. Fish Commission steamer *Albatross*, from which he and his crew routinely sampled with a plankton net.

Other early plankton studies were performed by biologists from Columbia University who booked quarters near Port Townsend in 1896 hoping to encounter species from both the protected waters to the south and the more open waters to the north, and who learned about oceanic waters by persuading a tugboat captain to take them out to Cape Flat-

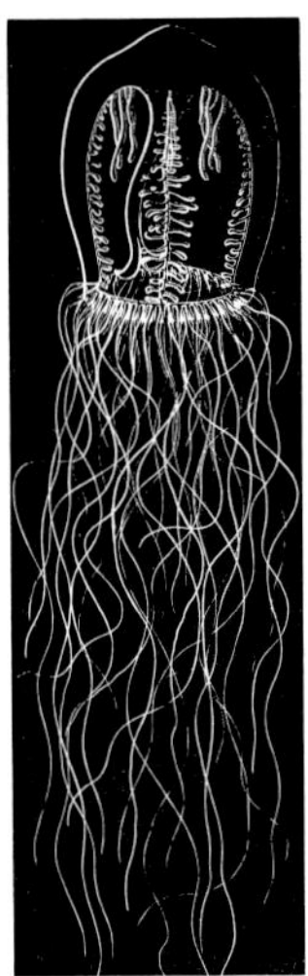

Figure 2.3 The medusa jellyfish *Polyorchis*, pictured in this woodblock print by Alexander Agassiz (1865), was among the first plankters studied on Puget Sound. The *Albatross*, one of the nation's first vessels built for marine research, visited Puget Sound near the turn of the century. (Photo courtesy Puget Sound Maritime Historical Society)

tery. They were visited by Trevor Kincaid, who later supervised the first indigenous plankton research on the Sound. The same group visited Port Townsend again in 1897, along with a contingent from England.

The Laboratory and the Field

With the founding of the University of Washington Friday Harbor Laboratory by Trevor Kincaid (Figure 2.4) and T. C. Frye biological oceanographic research on Puget Sound began to diverge into two channels: those studies carried out on location in the field (in this case, water), and studies in the laboratory. Laboratory and field studies differ in scope: there is a broad landscape focus in the field, versus a narrow portrait focus in the lab. Seagoing studies encounter all the complicated variables of ecosystems—light, temperature, weather, currents, and water chemistry—in a complex, intermeshing natural machinery. Field studies must integrate. Laboratory studies segregate, trimming problems down to size by eliminating or controlling as many variables as necessary or possible. Studying an ecosystem is like assembling a jigsaw puzzle. Field studies stand back to fit the pieces to a grand design, whereas laboratory studies move in close and fit pieces according to their shapes. Field studies sketch out a skeleton; laboratory studies put flesh on its bones.

An entire battery of subsciences has arisen from laboratory work. Perhaps the earliest laboratory work done at Friday Harbor was the study of morphology, the anatomy and external structure of organisms—shape, size, color, texture, and arrangement. When a new spe-

Figure 2.4 Trevor Kincaid, a precocious bug collector, was appointed a zoology professor at his University of Washington graduation. In 1904 he and botanist T. C. Frye founded the Puget Sound Biological Station (now the Friday Harbor Laboratories), where they pioneered research on local plankton and other marine life. Kincaid also founded the University of Washington Fisheries College in 1919 and helped revive the state's failing oyster industry in the 1920s. He died in 1970 at age 97. (Photo courtesy Friday Harbor Laboratories)

Figure 2.5 Thomas G. ("Tommy") Thompson was a University of Washington chemistry professor with an interest in seawater who in 1930 was appointed to transform the biological station at Friday Harbor into an oceanographic laboratory. His comprehensive and systematic research methods brought fresh and valuable insights into the physics, chemistry, and biology of Puget Sound, and he created a world-class oceanographic institution. He was the UW's first member of the National Academy of Sciences. After his death in 1961 the University's new ocean-going research vessel was christened in his honor. (Photo courtesy Friday Harbor Laboratories, University of Washington)

cies is identified (as many were at Friday Harbor), a full description of its morphology is published, in detail sufficient for other scientists to recognize the organism. Morphology is intimately bound up with taxonomy, the subscience of evolutionary and genetic classification.

Another early pursuit that led to an interest in plankton (especially the meroplankton) was embryology, the study of maturation from fertilized egg to maturity. The usefulness of this subscience had been demonstrated by J. Vaughn Thompson with his newly invented tow-net. A small zooplankter resembling a young crab was maintained in the laboratory and matured to become a barnacle. This dispelled the notion that the barnacle was a relative of the clam.

To maintain living organisms in the laboratory it was necessary to artificially duplicate natural living conditions. So began the study of physiology, the subscience of organisms' metabolic machinery. Phytoplankton can be grown just like a marine houseplant, with the proper temperature, light, salt, water, and fertilizer, although it took decades to find the correct recipe. The ability to culture phytoplankton made it much simpler to grow zooplankton in the laboratory as well, yielding knowledge of food preferences and requirements in the animals. Along the way, lab researchers discovered reactions of organisms to such unusual and stressful conditions as lack of food and extremes of temperature and salinity.

Field studies of plankton depend for their success on careful planning. The composition of the plankton can vary tremendously, so the location and timing of sample collection, as well as the equipment used, affect how well the samples represent the intended populations. Systematic field studies of Puget Sound plankton accelerated with the arrival at Friday Harbor of Thomas G. Thompson (Figure 2.5). During the decade from the mid-1920s to the mid-1930s, Thompson's research detailed plankton abundance in the Sound, using a logical and organized approach designed to illuminate the variability of plankton production at different times and places.

Thompson and his co-workers began their study by surveying the seasonal changes in the plankton from the pier at the Friday Harbor Lab for several years, comparing one year to the next. Their next step was to compare the plankton at Friday Harbor to that found at neighboring sites near Orcas Island and in the Strait of Juan de Fuca. Finally, spatial and temporal changes were charted on a multi-day cruise over the entire Sound. They observed significant differences in the plankton community over distances of a few kilometers horizontally and a few meters vertically. There were also radical changes at one location over the few hours it takes for the tide to change, over the few days it takes for the weather to change, and between the same date in different years.

Standing Stock and Productivity

To describe the changes in plankton abundance over time and by location, two fundamental questions need to be answered: "How much is present?" and "How quickly is it changing?" As a moving body is characterized by its position and its velocity, so a plankton community is described by a static quality, the standing stock, and a dynamic quantity, the rate of production per unit time, called the productivity, which will determine the future standing stock.

Standing Stock

Standing stock can be measured as either the number of organisms, the population, or their combined weight, the biomass. Neither by itself is a complete index of abundance, however, and no measurement of standing stock can be more reliable than the sampling methods from which it is determined.

The simplest method of assaying plankton standing stock, requiring only a microscope, is to count the population of organisms in a known volume of water. By microscopically counting plankters, the organisms can often be catalogued by genus and species, and even by age. The observer can judge the health of the organisms, and separate living organisms from dead ones. But counting has one disadvantage—it is tremendously slow, and thus quite expensive.

Biomass can be measured more rapidly than population, and by a less experienced observer. The simplest method is to measure the volume of settled plankton captured with a bottle or net. Inaccuracies due to the presence of water and other non-biological matter can be reduced by drying and weighing the sample, or by burning it and estimating organic matter from the amount of carbon dioxide driven off. The problem with any method for measuring biomass is the quality of the information it yields. There are no details of species present, no distinction between living and dead material, nor even a way to separate plants from animals.

The problem of separating plant from animal, and living cells from inert matter, has been at least partially solved, however. The plant pigment chlorophyll *a*, when routinely extracted and purified, has a distinctive green color that can be precisely sensed by an instrument called a spectrophotometer. The instrument detects the amount of pigment in the sample, and a simple correction can be made for the presence of non-living pigments. The method was developed at the University of Washington by Francis Richards with Thomas G. Thompson, and it has become an indispensable tool of the oceanographer's trade.

More recently, a faster and more sensitive (if less accurate) instrument has been developed for assaying chlorophyll: the fluorometer, which can detect the fluorescence of phytoplankters bombarded with light. This instrument can be combined with a pump, at some loss of sensitivity, to monitor the fluorescence of living phytoplankton in a continuous stream of seawater. It furnishes a powerful tool for rapid estimation of phytoplankton biomass over wide areas, like an EKG of the sea. Further promise is held by aerial and satellite photography.

There is great promise for developing an equally useful automated method for measuring zooplankton biomass, continuously and in the field. The method evolved before World War II when sailors noticed on their sonar depth soundings curious midwater blips and bands, which migrated up at night and down during the day. These echoes proved to be schools of fishes. When the frequency was stepped up to an inaudible 105 kilohertz, groups of large zooplankters could be detected. Like the fluorometer, the echo sounder unfortunately cannot distinguish one species of zooplankton from another. Together with the fluorometer, though, it might allow simultaneous acoustical and chemical surveys of great volumes of water for both plant and animal plankton. The information could be available almost immediately, without wetting a net.

Productivity

To transform a "snapshot" in time—as a standing stock measurement is—into at least a semblance of a moving picture requires some means of observing the rates of growth and reproduction of plankters, and ideally, their death rates as well. The rate of production, however, is more subtle and difficult to measure than standing stock. It requires prolonged observation and comparison of initial to final states. For this reason, methods for estimating growth rate and productivity have been slower to develop.

The simplest means of estimating productivity is to compare the standing stock before and after a fixed length of time. This yields net growth—the sum of birth, death, and migration. The accuracy of this method is likewise limited by that of the sampling methods. Small changes in standing stock due to production are likely to be invisible against the gross variations in the sea, and can be effectively monitored only under controlled laboratory conditions.

What is needed to measure productivity in the sea is a chemical whose concentration can be easily measured and has a direct relationship to growth, a chemical as characteristic of the growth process as chlorophyll *a* is characteristic of living plants. Fortunately, the unique process of photosynthesis, to which chlorophyll is vital, also provides a means by which the growth of plants (at least) may be monitored. The real business of plants is to "fix" carbon dioxide into glucose using sunlight. Carbon is the structural material and the energy currency of all biology. Measuring the rate of carbon fixation places a finger on the pulse of the phytoplankton, and of the entire ecosystem.

To measure phytoplankton productivity, a small bottle filled with seawater is injected with a shot of sodium bicarbonate (abundant in seawater, it generates carbon dioxide), in which there is an enriched proportion of radioactive carbon-14 atoms, the same isotope used for dating of fossils. The bottle is then suspended back in the water, or placed before a light of known intensity. As the algae photosynthesize, they take up the "hot" (radioactive) carbon along with the "cold" carbon already present in the water. After a few hours of incubation, the plankton is filtered, and on each filter the amount of radioactivity (above background) measured with a Geiger counter or similar device is proportional to the total amount of carbon fixed by the phytoplankton.

Unfortunately, there is no method comparable to the carbon-14 method to measure production by zooplankton, nor is there much prospect of one emerging soon. The rate at which animal standing stocks increase and decrease can be found only from consecutive biomass or population estimates, with all their imprecision. A newly developed method can, however, determine the amount of phytoplankton eaten, by assaying the fluorescence of chlorophyll in zooplankton guts.

Since World War II the study of plankton has undergone an explosion of youthful energy, during which time more was discovered than in the century before. The advent of chemical and automated methods marks planktology's scientific adolescence, but maturity is still far away. Similarly, oceanography as a whole is a superstructure assembled from other, more basic sciences, and cannot be expected to keep pace with them. As the 20th century wanes, oceanography could be compared to other sciences in their youth: medicine, for example, at the time of Louis Pasteur, or physics before Einstein. There is an understanding, but as yet no prediction or control, of major phenomena and their causes. The rate at which new knowledge emerges, while accelerating, is still slow. Textbooks of medicine are out of date almost as soon as they are printed, yet some oceanographers still use a text written forty years ago.

Biological oceanography is also a pure science. The applied sciences of fisheries and mariculture derive from it, just as horticulture and agriculture are applications of the pure science of botany. There are few practical applications to which oceanography can yet be put—we cannot yet farm the open sea or abate pollution, and we have little immediate prospect of doing either. Pure science, which inquires solely for the sake of inquiry, often seems to progress slowly, with no clear direction. A rapid advance in applied science, however, is but the tip of an iceberg buoyed by an unseen mass of pure science. Future husbandry of the seas will depend on today's primitive inquiries, as space travel was begun by Newton.

CHAPTER THREE

Plankton Hall of Fame

Natura nusquam magis est tota quam in minimis (Nature is to be
found in her entirety nowhere more than in her smallest creatures)
Pliny

A community of similar plankton species inhabits much of the Pacific coast of North America, from California's Golden Gate to Alaska's Inside Passage. This community is exemplified by the plentiful plankters living near the center of that range, in the sheltered and beneficent waters of Puget Sound. Puget Sound is to plankton what Florida is to oranges, what Iowa is to corn, what the Cascades are to the Douglas fir. Although hundreds of species of plankton may populate the Sound at one time, or over the course of a year, most of the plankton assemblage is comprised of a relatively few important organisms.

Phytoplankton

The species of algae which comprise the marine phytoplankton are unique life forms. Although among the oldest forms of life on the planet, they are also highly specialized and well-adapted to their environment. Unlike higher plants, which can contain billions of cells, each phytoplankter consists of only a single cell. Though they sometimes link together to form chains, each individual cell is self-sufficient; the chains are colonies, not organisms.

What distinguishes an organism as a plant is the process of photosynthesis. Plants capture the energy of the sun in the green pigment chlorophyll, and use this energy to convert carbon dioxide and water into sugar. Phytoplankters do this only within the proper range of light intensity, temperature, and salinity conditions. In addition, as a garden needs fertilizer, phytoplankters need certain nutrients. The required nutrients are nitrogen, most common in the sea combined with oxygen as nitrate; phosphorus, present as phosphate; carbon, potassium, sodium, calcium, and sulfur, all plentiful in the sea; and trace metals such as iron, manganese, cobalt, copper, nickel, tin and zinc. A majority of phytoplankters also need vitamins, especially vitamin B_{12}.

According to the "limiting nutrient" concept derived from terrestrial agriculture, the supply of these plant nutrients limits both the productivity and the standing stock of phytoplankton. Nutrients are present in seawater in roughly constant ratios, which do not necessarily match the demands of phytoplankton, the theory continues. One nutrient will be exhausted before the others, and it—the limiting nutrient—will place a ceiling on plant growth. The limiting nutrient in fresh water is usually assumed to be phosphorus. In salt water, however, nitrogen is believed to be in shorter supply, sometimes consumed by phytoplankton to levels of virtual undetectability at the sea surface.

Nitrogen supply is only one influence on phytoplankton growth in Puget Sound, and is believed to be less important in most instances than the availability of light. The effects of the trace nutrients, such as minerals and vitamins, have received little study and may also have an influence. Furthermore, when the complexities of multiple phytoplankton (and zooplankton) species and multiple nutrient sources in a turbulent fluid are considered, the limiting nutrient theory can play only a partial role in controlling phytoplankton abundance in the Sound.

All phytoplankton cells have an outer cell membrane and an inner, jelly-like cytoplasm. Many also possess an armored cell wall. All reproduce asexually by simple cell division, called binary fission. A parent cell expands its size, duplicates all its intracellular contents, forms a wall through its midsection, then separates into two sibling cells. The siblings are genetic duplicates (clones) of each other, and the parent ceases to exist. Binary fission is a powerful process: it expands the population by doubling again and again, creating from a single cell first two, then four, eight, and sixteen cells. Propagating in this fashion, once a day, a single cell can produce a billion copies of itself in a month. Binary fission is the means by which the cells of all larger organisms reproduce. Only in unicellular organisms such as phytoplankton, however, do single divisions produce complete offspring.

Phytoplankters can also undergo sexual reproduction, in which a cell dissolves into tiny, swimming sperms or eggs, which unite in the water to form zygotes. The new cells that form are genetic crosses between the parents, rather than duplicates of a single parent. Sexual reproduction has received little study, but is believed to be triggered by a sudden deterioration in such environmental conditions as light intensity, temperature, or nutrient concentrations. The exact cues are likely to be different for each species.

An unfavorable environment may also trigger the formation of any of a variety of resting spores or cysts, with hardened walls and slowed metabolism. These sink to the bottom where they are less likely to be eaten, then regenerate vegetative cells when favorable conditions return, possibly months or years later. Cyst formation is poorly understood and may be associated with the sexual cycle. It is observed only in species inhabiting coastal waters sufficiently shallow that storms and currents can resuspend bottom sediments and cysts into lighted waters where growth can resume. The triggering of phases of the sexual and cyst cycles simultaneously in entire populations of a phytoplankton species is suspected of accounting for very rapid shifts in phytoplankton species composition occasionally observed in Puget Sound.

Of the many diverse phytoplankters in the sea, those most common in Puget Sound can be lumped into three broad categories. The customary unit for such classifications of both plants and animals is the phylum, the coarsest taxonomic division after the kingdom. Below the rank of phylum, in order of increasing specificity, are class, order, family, genus, species, and race or strain. Scientists can debate endlessly over such distinctions, however, and for our purposes it is more meaningful to define pragmatic ecologically based groupings, which cut across strict phyletic lines. In addition to diagnostic external features such as size and cell wall structure, the most useful characteristic for differentiating the three groups of phytoplankton in Puget Sound is locomotion, which is found in the plant kingdom only in certain microalgae.

The possession of one or more whip-like swimming appendages called flagella is common to many groups of unicellular organisms not far from the fork in the road of evolution separating plants and animals. These organisms are called flagellates, and the phytosynthetic forms are called phytoflagellates. Their swimming speed—at most a few meters per hour—does not threaten their status as plankton. The dinoflagellates, including both plant and animal forms, have two uniquely arranged flagella and a distinctive cell shape. The term phytoflagellates will be used below to refer only to the remainder of the phytosynthetic flagellates, a heterogeneous assortment sharing similar size and cell wall characteristics.

The diatoms have no flagella (except on their male reproductive cells), and therefore planktonic diatoms have little if any locomotion. Other species of non-planktonic diatoms important in shallow benthic environments have a snail-like gliding ability not involving flagella. Varying widely in size, diatoms also have a unique hardened cell wall.

Diatoms (Bacillariophyceae)

Diatoms (Figure 3.1) are widely assumed to be the most common plants in the sea, although new findings on flagellates may challenge this belief. The name denotes the presence of two hard outer shells, called frustules, embedded in the wall around each cell. Shaped like two pillbox halves that fit one rim inside the other, the frustules are made of silica extracted from seawater, making silicon an additional essential nutrient for the diatoms. The casing is perforated, allowing chemical exchange between the cell and the water, and is sculpted in each species into a different exquisite pattern. Because new frustules form within those of the parent during binary fission, the cells of a diatom population steadily diminish in size during asexual propagation. Only sexual reproduction can restore the original cell size and vigor of the population; otherwise the cells die out.

 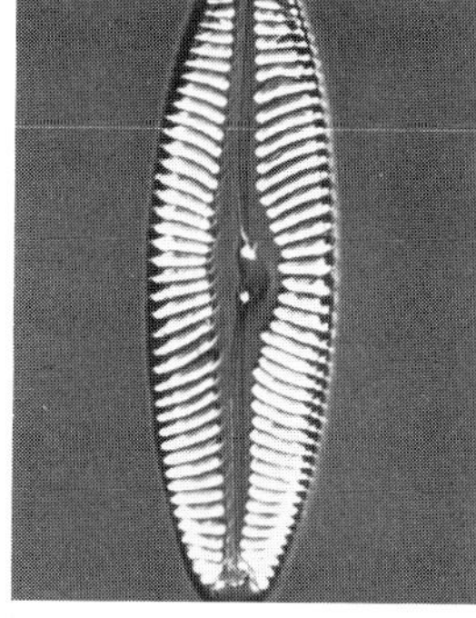

Figure 3.1 Diatoms

Top left: Separated pieces of the frustule of the solitary centric diatom *Coscinodiscus*. Actual diameter approximately 100 micrometers. (Courtesy National Marine Fisheries Service (NMFS), NOAA, micrograph by Michael Eng).

Top right: The solitary pennate diatom *Navicula*. Actual length approximately 100 micrometers. (Courtesy School of Oceanography, University of Washington)

Bottom left: A portion of a chain of the centric diatom *Skeletonema*. Actual cell diameter approximately 25 micrometers. (Courtesy B. Dumbauld)

Bottom right: A segment of the centric diatom *Chaetoceros*. Actual cell diameter approximately 35 micrometers. (Courtesy Beatrice C. Booth)

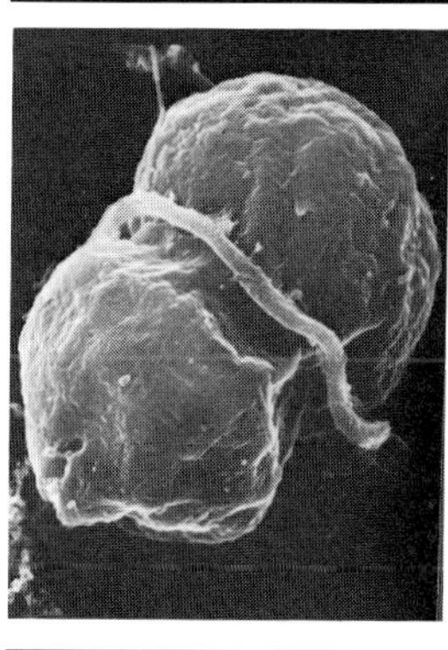 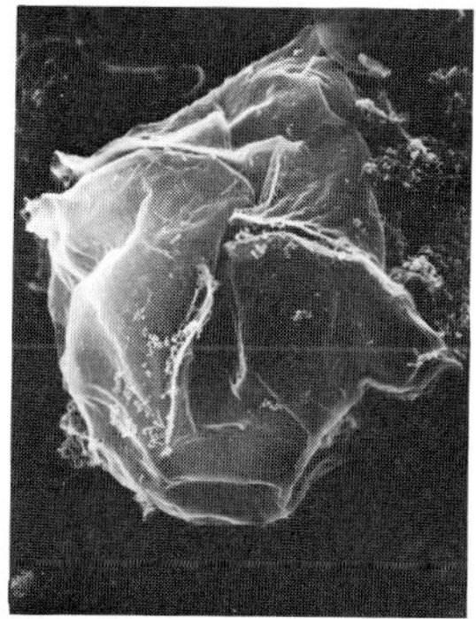

Figure 3.2 Dinoflagellates

Left: The unarmored dinoflagellate *Gymnodinium*. Note the flagellum in the transverse groove. Actual cell length approximately 6 micrometers.

Right: The armored dinoflagellate *Gonyaulax*, cause of paralytic shellfish poisoning. Actual size approximately 30 micrometers. (Photos courtesy Susan B. Stanton)

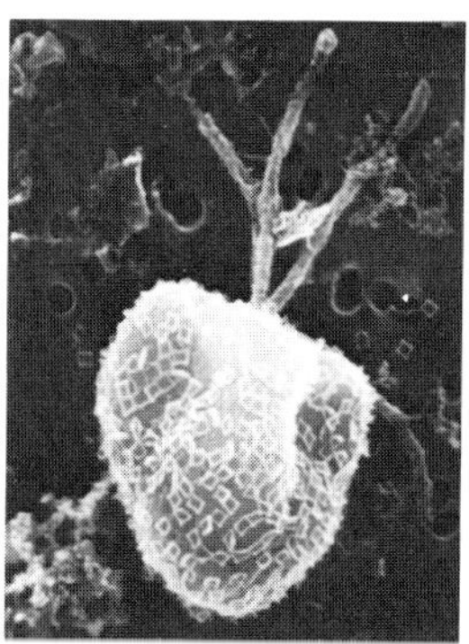 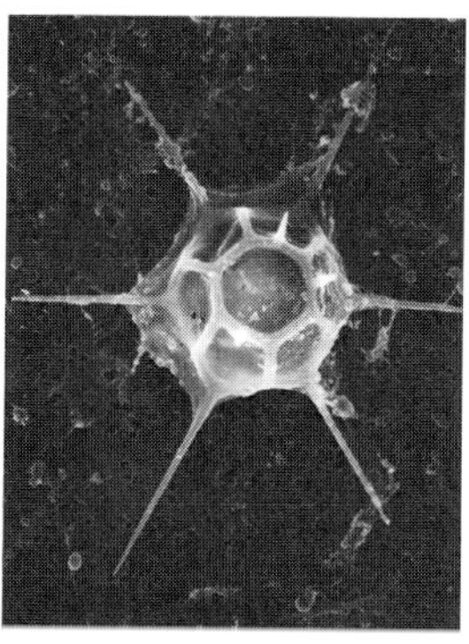

Figure 3.3 Phytoflagellates

Left: A green flagellate (probably *Pyramimonas*). Note flagella and surface scales. Actual size approximately 8 micrometers.

Right: The skeleton of the golden-brown silicoflagellate *Dichtyocha*. Actual size including spines approximately 50 micrometers. (Photos courtesy Beatrice C. Booth)

The opaline diatom frustules are a major constituent of marine sediments at high latitudes, since they are heavy and slow to redissolve as they sink. Useful as fossil indicators, they compose the fine mineral diatomaceous earth (diatomite), which because of its large surface area and abrasiveness has a wide variety of commercial uses, such as in wine filtration and in scouring powder. Diatomite deposited before the Cascade Mountains rose from the sea has been mined in the Columbia River basin near Vantage, Washington.

Two classes of diatoms, centric and pennate, are distinguished by shape. Centric diatoms are radially symmetrical like a wheel. Pennate diatoms are bilaterally symmetrical or asymmetrical, and elongated along one axis like an almond. Both classes contain some species which are solitary and others that link together to form chains. Long spines may also be present in some colonial species.

Dinoflagellates (Dinophyceae)

The dinoflagellates (Figure 3.2) are a highly contradictory group about which it is difficult to make generalizations. Most are plants, but some are animals, parasites, or other innovative forms. The prefix "dino-" (as in dinosaur) refers to a cell wall armored with cellulose plates, but many species are naked. Most dinoflagellates have a basic body plan resembling two cones joined at their bases, with a groove in between. One flagellum lies in the groove and its beating makes the cell spin like a top. The other flagellum lies in a second, perpendicular groove on the lower cone and its beating propels the cell forward. A major subgroup of dinoflagellates, however, is almond-shaped, with both flagella inserted at one end. Some dinoflagellates, like some diatoms, drastically modify their basic body plan by forming chains, or (more commonly) by having prominent spines.

Dinoflagellates have a peculiar trait of migrating vertically in the water. During the daytime, they are found near the surface photosynthesizing. In the evening, however, they may be found a few meters below the surface, and as dawn approaches they swim back to the surface en masse. They form dense swarms at times and places where conditions are right, resulting in a "red tide." Some dinoflagellates are luminescent and are one cause of the glow of stirred water on summer evenings, and some can sicken or kill marine animals and even humans.

Phytoflagellates

The most common species of the other phytoflagellates (also sometimes called nanoflagellates) in the Puget Sound area are mostly unarmored, usually unicellular, come from many different taxonomic groups, and are distinguished mainly by being the smallest of the phytoplankton (Figure 3.3). They have been overlooked for many years because their size makes them difficult to capture and preserve. Little is yet known of their habits or abundance in Puget Sound. Through careful sampling, however, phytoflagellates have been discovered in a wide variety of habitats, and may prove to be very important.

Zooplankton

The zooplankton is an extremely diverse group of animals, containing at least one developmental stage of dozens of phyla. Zooplankton lifestyles are far more varied than those of phytoplankton. Animals must have senses to detect, propulsion to pursue, appendages to apprehend, and mechanisms to ingest, digest, and egest food. Animals also use sophisticated strategies to avoid becoming food. Some zooplankters, the protozoans, are unicellular, but most are multicellular metazoans. Animal reproduction, especially that of the metazoans, is more complex than that of plants. In most cases, asexual reproduction has been replaced by sexual reproduction, and more advanced animals have prolonged stages of immaturity and more carefully programmed behavior for bearing young to increase their odds for survival.

Protozoans

Most of the unicellular members of the phylum Protozoa (Figure 3.4) reproduce as phytoplankters do, by simple binary fission. Most protozoans also feed on very small phytoplankton, especially phytoflagellates, or on bacteria. Four types of protozoans are encountered in Puget Sound: foraminiferans and radiolarians are uncommon, and dinoflagellates and ciliates are more abundant.

The mouthless foraminiferans (forams) and radiolarians eat by simply engulfing their food, as an amoeba does. Both possess skeletons: foram shells are external, calcareous, and whorled like those of a tiny snail, while radiolarians have star-shaped inner silica skeletons. Both types of skeleton form thick oozes over wide areas of ocean bottom. The white cliffs of Dover are 500-foot-thick piles of ancient forams.

There are few common or well-studied zooplanktonic flagellates in Puget Sound. The most prominent is an animal dinoflagellate, *Noctiluca*. Mostly water like a jellyfish, *Noctiluca* is unarmored and is either transparent or pigmented a faint pink. It has one flagellum for swimming, while the other, next to its simple mouth, is used as a tentacle for grasping food. It can reproduce rapidly enough to form a luminescent animal red tide, and it is a voracious feeder on all types of phytoplankton and even on small animals, some of which can be seen inside its diaphanous body before they are digested.

The ciliates, the most consistently abundant protozoans in Puget Sound, derive their name from the rows of tiny hair-like cilia that they use for both propulsion and food-gathering. Some ciliates, in addition, have forms of armor. The tintinnids have a tough outer shell called a lorica, always a variation of the basic ice-cream-cone shape, which they adorn with cemented bits of sand. When threatened, a tintinnid will abandon its lorica. Less armored than the tintinnids are the oligotrichs, which cannot abandon their sheaths and sometimes have none at all. Holotrich ciliates are represented in Puget Sound principally by one species, *Mesodinium rubrum*, which can at times form red tides. *Mesodinium* has only a vestige of a mouth, and instead of eating lives off symbiotic phytoflagellates within its body; it is an animal which has reverted to a plant-like existence.

Crustaceans

The class Crustacea (Figure 3.5) of the phylum Arthropoda clearly dominates the zooplankton of Puget Sound and the sea as a whole. The marine equivalent of the insects (another Arthropod class, which out-numbers all other animal species combined) is an order of crustaceans called the copepods. Copepods are superfically shrimp-like, with seg-mented, torpedo-shaped bodies, antennae and mouth parts at the front, and swimming appendages dangling below. Corkett and McLaren have surmised that the copepod *Pseudocalanus* may be the most populous metazoan genus in the world, and Hardy has ventured that the number of copepods may exceed the numbers of all other metazoans in the sea combined.

Most copepods have preferences for either plant or animal food, and their sizes and feeding structures differ accordingly. *Calanus*, a moderate-sized animal that can contribute a large fraction of the cope-pod biomass in Puget Sound, is a grazer that as an adult eats primarily diatoms, although given the opportunity it will prey on some small pro-tozoans and larvae. While its precise feeding mechanism is not yet clear, *Calanus* appears to draw water toward its mouth using feeding appendages with fine branches, which concentrate masses of small food particles as a phytoplankton net does. This type of strategy is called suspension feeding.

Contrasting with *Calanus* are some strictly carnivorous copepods. The largest copepod in the Sound, about the size of a grain of rice, is *Euchaeta*, which finds the smaller herbivore *Pseudocalanus* an ideal prey. A unique predator is the smaller *Corycaeus*, often the most nu-merous in the Sound. It has sharp, pincer-like appendages for grasping individual prey, and even has primitive eyes. Such carnivorous cope-pods also have blunt, molar-like grinding surface on their jaws for chewing muscle, while more herbivorous animals such as *Calanus* have sharp, incisor-like jaws for cracking open diatom frustules.

Crustaceans rely exclusively on sexual reproduction. A few hours after copepod eggs are laid, they hatch into larvae called nauplii, which swim and like the protozoans feed on the tiniest phytoplankton. In copepods, the larva molts eleven times before reaching adulthood, passing through an additional series of copepodite stages. The entire life cycle, depending on temperature and the availability of food, takes about a month in the summer, or several months in the winter.

Members of orders of larger shrimp-like planktonic crustaceans—the euphausiids, mysids, and hyperiid amphipods—are second in abundance to the copepods. In addition, one true shrimp, the small *Pasiphaea*, is planktonic; but the Puget Sound commercial and sport shrimp *Pandalus* is too large to be a plankter except in its larval stages. Euphausiids, mysids, and amphipods live deeper in the water than many of the copepods, feeding on the largest phytoplankton, protozoa, copepods, and other medium-sized zooplankters including fish larvae. Many of them also migrate a hundred meters or more below the surface during the daytime. The euphausiids, especially the genus *Euphausia*, are "krill," the staple diet of baleen whales in the Arctic and the Antarctic. These larger and more complex organisms pass through more stages of immaturity than the copepods. Euphausiids usually take an entire year to mature as plantonic larvae, and live for two years or more. Larval mysids and amphipods, in contrast, are reared in a kangaroo-like maternal brood pouch before becoming free-living.

Two minor orders of crustaceans, both of which are more common in fresh water, are the ostracods and the cladocerans. Ostracods have a shell shaped into two clamshell-like cups, in which they hide as they float, emerging only to swipe at passing food. Cladocerans, the marine relatives of *Daphnia* (the water-flea of lakes and ponds), have a prominent eye-spot, and long limbs used for both swimming and feeding.

Rotifers

The phylum Rotifera (Figure 3.6) includes mostly freshwater members, but the genus *Synchaeta* appears regularly in Puget Sound. Rotifers are multicellular, but similar in size to the larger protozoans. They feed on large phytoplankton and small animals, using a ring of beating cilia which surrounds powerful jaws.

Coelenterates

The phylum Coelenterata, also called Cnidaria, (Figure 3.7) is best known, perhaps, for corals and anemones. Its local planktonic representatives are species of floating or swimming jellyfishes, called medusae. The medusa is but one stage of the animal's life cycle, alternating with a sedentary form called the hydra. For many years the medusae and hydras were thought to be different species entirely, but through slow, painstaking observation the matching life stages were pieced together. Hydras are asexual and bud off large numbers of medu-

Figure 3.4 Protozoa

Left: The dinoflagellate protozoan *Noctiluca* that causes nontoxic luminescent red tides in Puget Sound. Note the flagellum. Actual diameter approximately 100 micrometers.

Right: The lorica of the ciliate protozoan *Tintinnopsis*. Actual length approximately 100 micrometers. (Photos courtesy Alexander J. Chester)

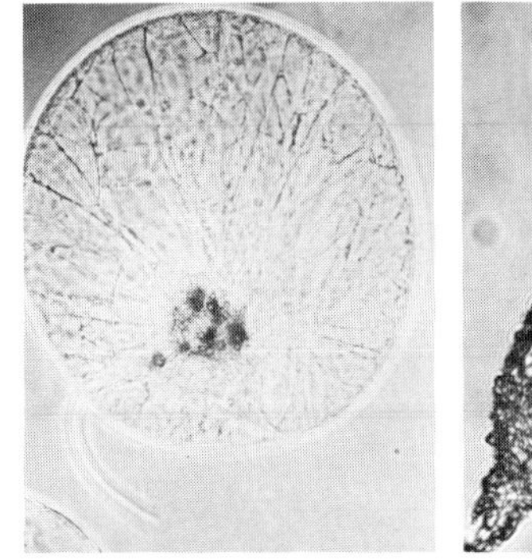
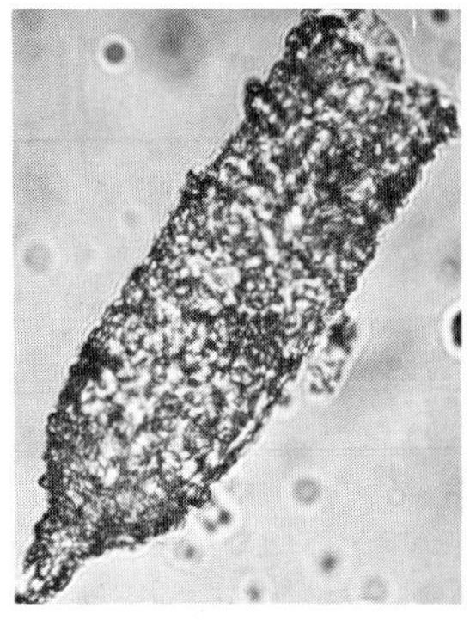

Figure 3.5 Crustaceans

Left: The herbivorous copepod *Calanus* showing the antennae and feeding (upper) and swimming (lower) appendages. Actual length approximately 3 millimeters.

Right: The carnivorous copepod *Euchaeta*, with coarse feeding appendages adapted for grasping, in contrast to the filtering apparatus of *Calanus*. Actual length approximately 1 centimeter. (Photos courtesy Charles H. Greene)

The euphausiid "krill" *Euphausia*, which is mostly herbivorous in Puget Sound. Actual length approximately 2 centimeters. (Courtesy Mark D. Ohman)

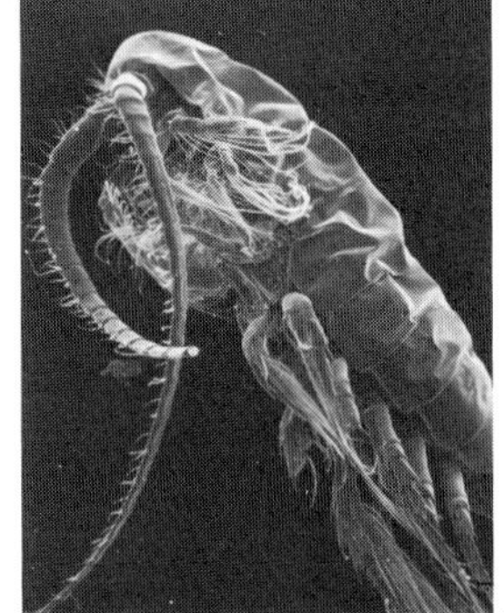
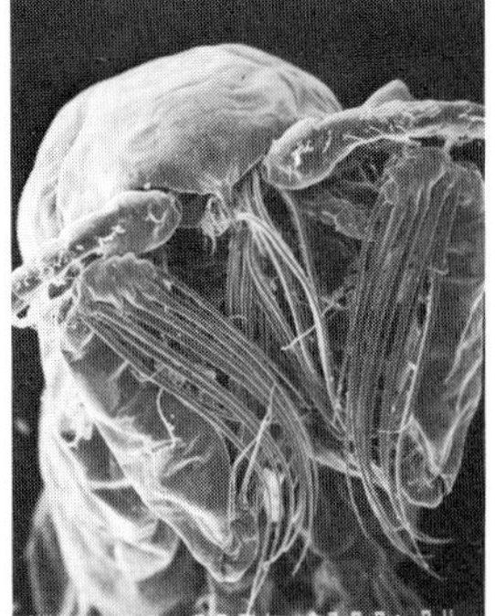
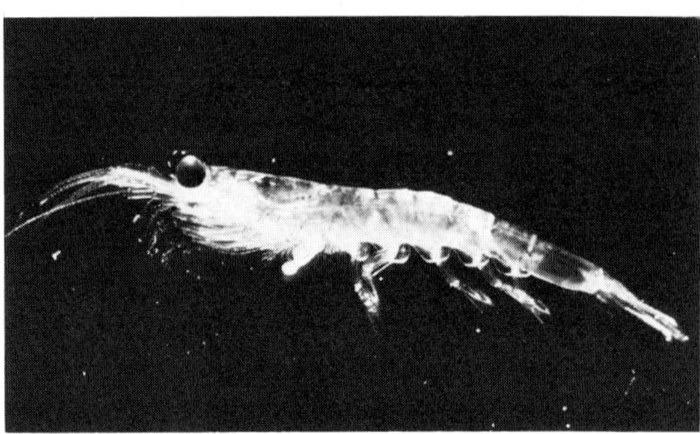

Figure 3.6 Rotifers

The brackish-water rotifer *Brachionus*. The blur is caused by beating cilia. Actual length approximately 200 micrometers. (Courtesy Richard Kaiser)

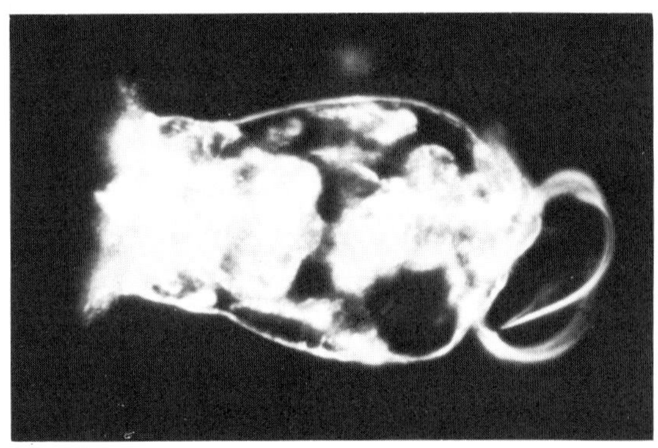

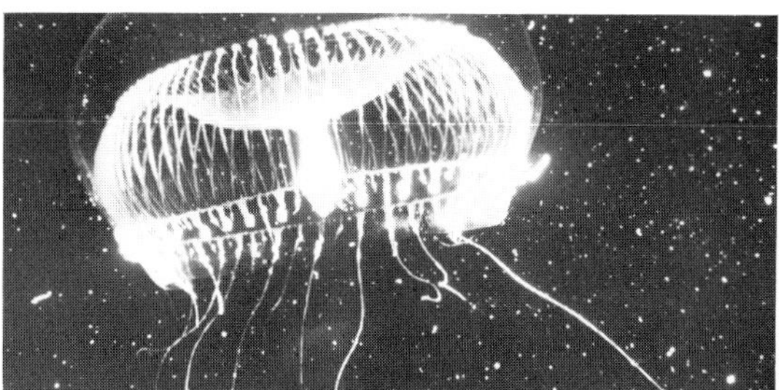

Figure 3.7 Coelenterates

The luminescent medusa stage of *Aequorea*. Actual size approximately 7 centimeters. (Courtesy William D. Waddington)

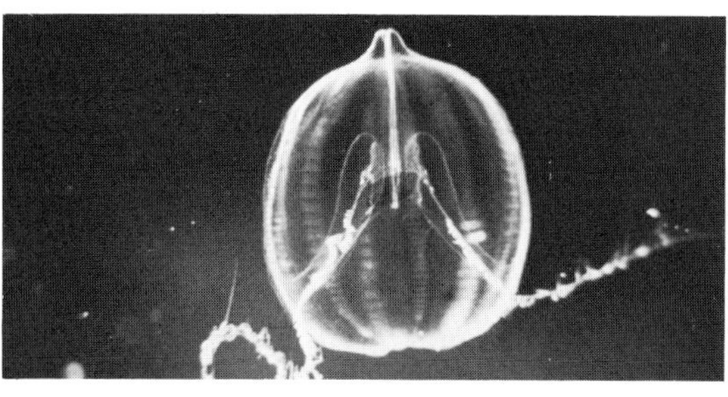

Figure 3.8 Ctenophores

The ctenophore *Pleurobrachia*. Note the comb rows and tentacles. Actual diameter approximately 3 millimeters. (Photo courtesy School of Oceanography, University of Washington)

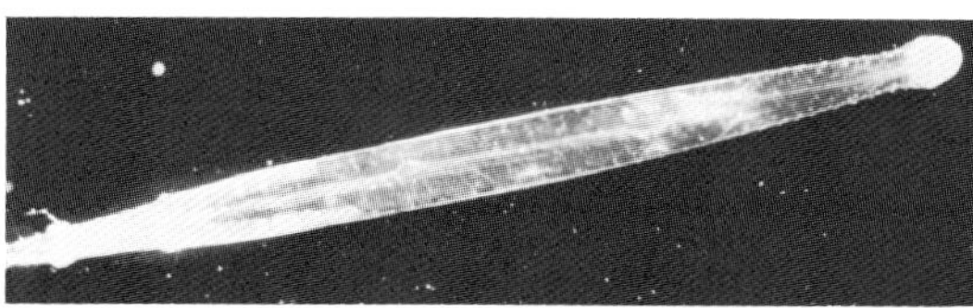

Figure 3.9 Chaetognaths

The chaetognath *Sagitta*. Spines at head are retracted. Actual length approximately 2 centimeters. (Courtesy Mark Ohman)

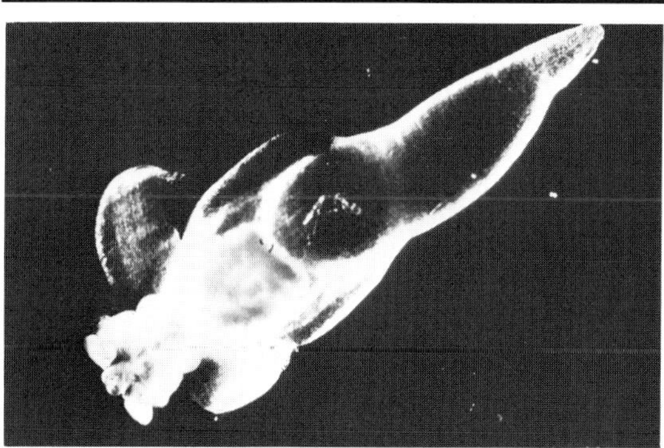

Figure 3.10 Molluscs

The naked pteropod *Clione*. Note "wings" (actually the "foot"). Actual length approximately 3 centimeters. (Photo courtesy School of Oceanography, University of Washington)

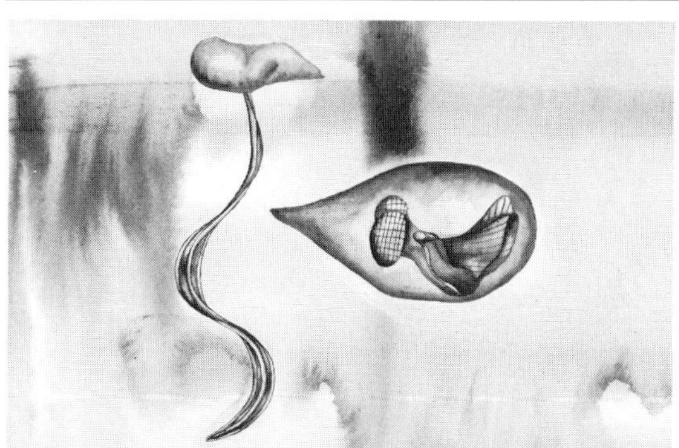

Figure 3.11 Chordates

The larvacean *Oikopleura*. The animal's head is at center. Its waving tail draws water through the surrounding mucous "house" to catch food on two sets of screens. Actual length including house approximately 2 centimeters.

sae, which carry out sexual reproduction as they drift. Coelenterates feed by using special organelles called cnidoblasts, which eject a thin cord tipped with a poisoned dart to embed itself in the prey. At times in the summer, medusae can float by the thousands near the surface and decimate the populations of their zooplankton prey.

Ctenophores

The phylum Ctenophora (Figure 3.8) was long included in the Coelenterata because its members have soft, transparent bodies and tentacles. Their delicate form makes them difficult to sample and preserve, so less is known about them than about most zooplankters. Ctenophores are nicknamed "comb jellies" for the eight rows of paddle-like fused cilia which divide their spherical bodies like the sections of an orange. These comb rows, which may be brightly luminescent, are the animal's means of propulsion. *Pleurobrachia*, the dominant ctenophore on Puget Sound, also has a long pair of tentacles with adhesive cells for snaring zooplankton prey, but it lacks the "stingers" of the coelenterates. Ctenophores have a tremendous appetite—up to tens of times their body weight per day—and when food is plentiful they can form dense, glowing swarms at the water surface.

Chaetognaths

The principal genus of the phylum Chaetognatha in Puget Sound is *Sagitta* (Figure 3.9), which can reach a length of two centimeters. The nickname "arrow worms" comes from the long, streamlined bodies and the feeding habits of the chaetognaths, which resemble those of a barracuda. Chaetognaths are voracious predators on all other zooplankters and larvae, even on some animals larger than themselves. They dart at high speed, snatch their prey with a lightning-quick strike, and swallow it whole. The chaetognaths, as well as the ctenophores, are hermaphroditic.

Molluscs

Like the Coelenterata, the phylum Mollusca is better known for its non-planktonic representatives: squid, octopi, clams, mussels, snails, and slugs (Figure 3.10). The nudibranch snails, known for their beauty, are sometimes found floating freely in the water but are dependent on rooted subsurface vegetation and attached animals, and so are not plankton. The truly planktonic molluscs, nicknamed "sea butterflies," are the pteropods. They include the shelled *Limacina* and the naked *Clione*. *Limacina* leads a life like other tiny shelled plankton; it has little swimming power and is a passive filterer of phytoplankton. It captures food by secreting a huge parachute of mucous, free-falling through the water, then eating the catch, mucous and all. *Clione* is a large carnivore (up to seven centimeters), which swims by making water wings of the flaps of skin that form the "foot" in its cousin, the snail.

Chordates

Some plankters belong to the same phylum as humans, the phylum Chordata (Figure 3.11), which includes the subphylum Vertebrata. In addition to the vertebrate fish larvae that exist in the plankton, there are adult organisms, such as the larvacean *Oikopleura*. It is a small, worm-like creature, which secretes a balloon-like, luminescent, mucous "house" that doubles as a filter for capturing phytoplankton. *Oikopleura* waves its tail to propel a stream of water through two sets of windows in the house, where food is caught on screens. When the screens clog, the animal simply abandons its house and erects a new one, sometimes as often as every four hours. The filter is fine enough to capture even the smallest phytoplankton and bacteria, making *Oikopleura* perhaps the only animal of its size to do so. It is a hermaphrodite, and a newborn can have its own house built and functioning within one day.

Planktonic Larvae

The animals above, except the medusae, spend their entire lives, including any stages of immaturity, in the plankton. About three-fourths of all the remaining animals in Puget Sound (nekton and benthos) have meroplanktonic larvae (Figure 3.12). At the surface the young find the food more abundant, the water warmer, and the currents for transporting them to new territory stronger than in most subsurface habitats. A piece of bare rock or wood exposed to the surface waters of Puget Sound will, within a matter of days, have a thin glaze of larval colonists: shipworms, barnacles, mussels, bryozoans, sponges, and other creatures. Many other larvae settle to the bottom in deeper water, responding to subtle chemical and biological cues. Millions of these ciliated floaters are released to ensure that a few survive. Although an animal might spawn but once a year, and spend only a few days in the plankton, in aggregate the planktonic larvae form an important component of the zooplankton from spring through fall. Although crustaceans are probably most numerous, many larvae are also contributed by animal groups such as the annelids, the molluscs, and the echinoderms.

Annelids are segmented worms that live on or in the bottom. Their larva is called a trochophore. Molluscs have a veliger larva with two wing-like lobes resembling those of an adult pteropod. Echinoderms, a highly specialized phylum that includes the sea urchin, starfish, sand dollar, and sea cucumber, also have a variety of highly specialized larvae. All of these larvae, while present in the plankton, occupy a niche similar to that of a large protozoan or a small copepod, feeding on phytoplankton near the surface.

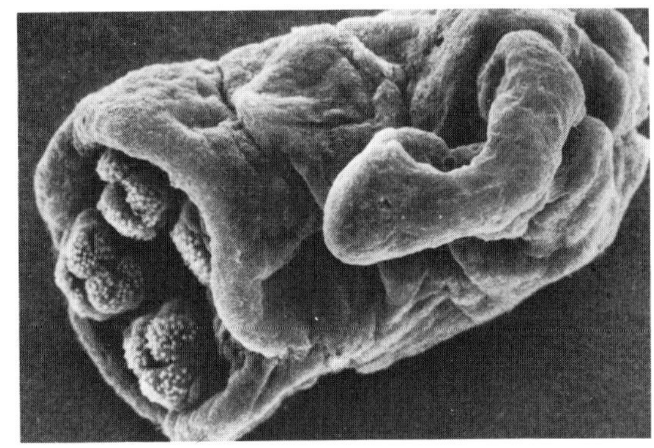

Figure 3.12 Larvae

A larval sea cucumber. Actual length approximately 0.8 millimeters. (Courtesy NMFS, NOAA, micrograph by Carla Stehr)

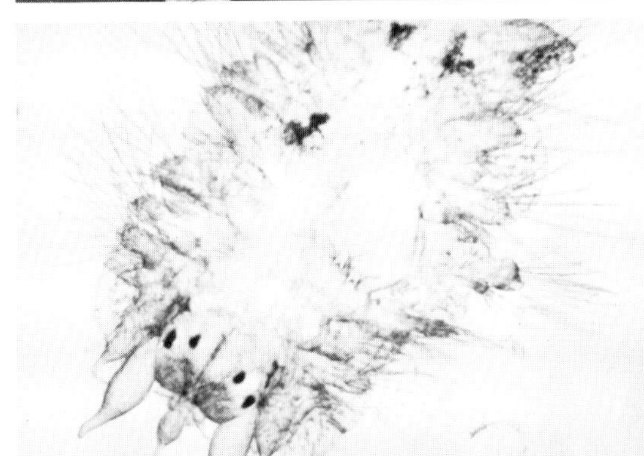

A larval annelid (polychaete) worm. Actual diameter approximately 0.5 millimeters. (Courtesy Alexander J. Chester)

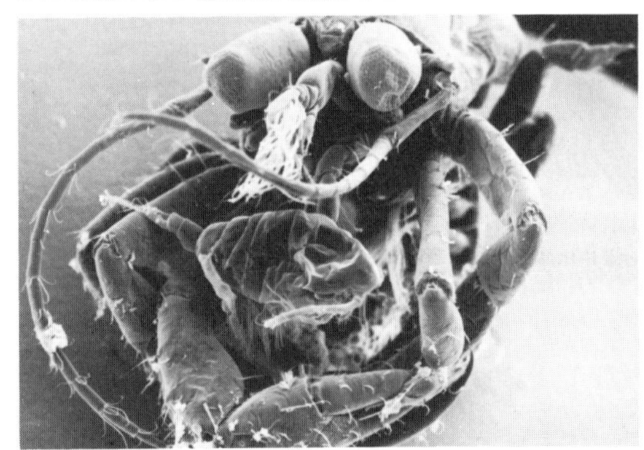

A larval hermit crab, with a small copepod (foreground). Actual length approximately 6 millimeters. (Courtesy NMFS, NOAA, micrograph by Carla Stehr)

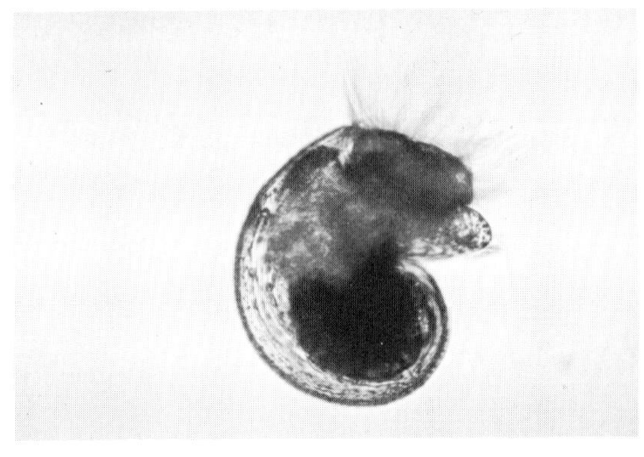

A late-stage larval gastropod mollusc. Actual diameter approximately 0.5 millimeters. (Courtesy Alexander J. Chester)

Larvae of such well-known nekton as smelt, herring, cod, rockfish, greenling, and flatfish also can be considered zooplankton during their earliest stages. Called ichthyoplankton, they are most common during winter and spring. They can be extremely abundant, and are important as a critical phase in the lives of economically valuable fishes.

This assemblage of planktonic organisms, with its diverse lineage, also exhibits a great variety of survival strategies. In part these strategies reflect the fundamental ecological duties of food production by phytoplankton, and consumption of and competition for that food by zooplankton. (These two roles will be examined in Chapters Five and Six.) But plankton must also contend with the unique environment in which it is immersed. The fluid nature of Puget Sound not only provokes many of the innovative adaptations of individual plankters, it even dictates the unique ways in which these organisms interact.

Logarithmic Size Comparisons	Meters	Object
	10^{-6}	Bacteria
	10^{-5}	Phytoflagellate
	10^{-4}	Diatom
	10^{-3} (1mm)	Copepod
	10^{-2} (1cm)	Euphausiid
	10^{-1}	Herring
	1	Salmon
	10^{1}	Killer Whale
	10^{2}	Super Ferry
	10^{3} (1km)	Tacoma Narrows Bridge
	10^{4}	Mt. Rainier
	10^{5}	Puget Sound

Figure 4.1 The lengths of common Puget Sound organisms and objects are compared here on a scale of powers of ten. Each object pictured is ten times the size of the one above it, and a tenth the size of the one below it. The euphausiid is shown actual size, 2 centimeters.

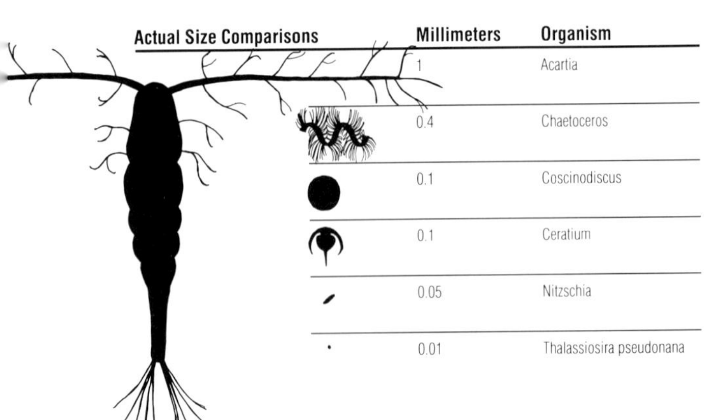

Actual Size Comparisons	Millimeters	Organism
	1	Acartia
	0.4	Chaetoceros
	0.1	Coscinodiscus
	0.1	Ceratium
	0.05	Nitzschia
	0.01	Thalassiosira pseudonana

Figure 4.2 The size of a small copepod is compared to that of typical phytoplankton cells, all enlarged about forty times.

CHAPTER FOUR

Seascapes

As soon as you have entered this pelagic wonderland, you will see that you cannot leave it.

Johannes Müller

Puget Sound, like the sea at large, is a wilderness at our doorstep, in which no people live and through which no trails pass. Most people who see or traverse Puget Sound know little of its creatures or their lives below the surface. To learn the inner workings of the ocean requires an imaginary journey like that of Lewis Carroll's Alice. We must step through the looking-glass surface of the sea, become very small, and see life from a plankter's point of view. Doing this, we indeed enter a wonderland.

The world of the plankton is alien to our own. Underwater life constrains plants and animals into adaptations and lifestyles very different from those on land. This is true not only in individual organisms, but also in the fundamental structure of underwater ecosystems. To understand what makes Puget Sound such a productive place, we must first look at some of those differences.

Size

Let us begin by putting oceanic dimensions in perspective. Puget Sound, at its deepest point, was scoured by glaciers to a depth of 275 meters; its bathymetry essentially mirrors the surrounding foothills. As on land, this relatively small departure from sea level produces dramatic physical, chemical, geological, and biological effects. The size of the environment is, however, not so important as the size of the plankton itself—this is the greatest difference between terrestrial and aquatic ecosystems. In the sea there are many advantages to being small, and few to being large. Although the largest animal ever to have lived—the blue whale—lives in the sea, on the whole marine organisms are much smaller than those on land (Figure 4.1).

Phytoplankton cells, from the smallest phytoflagellate to the largest diatom, range from about 5 to 200 micrometers (millionths of a meter) in diameter (Figure 4.2). Some diatoms chains can reach 500 micrometers in length. Phytoplankters fall into two size classes, the nanoplankton (less than 20 micrometers in diameter), and the net plankton (greater than 20 micrometers and large enough to be caught with nets). The phytoplankton spans a range of one millionfold in weight—the same range that encompasses all terrestrial mammals, from mouse to elephant. It is difficult, however, to grasp how small single cells really are. It would take about one trillion large phytoplankton cells, 100 micrometers across, to occupy the same volume as an average six-foot man. The same number of humans could be stacked into a cone the volume of Mt. Rainier. Plankton inhabits a Lilliputian world that even the power of the microscope can hardly make familiar.

The phytoplankton world, furthermore, is one of evanescence. Due to the simplicity and homogeneity of its environment, a phytoplankter has no need of roots, stems, trunks, branches, leaves, or flowers. The adaptations of the phytoplankton are simple: small size to absorb nutrients and retard sinking; spines to slow sinking and to discourage hungry animals; and the ability to propagate rapidly when conditions are favorable, as well as the tenacity to endure when they are not. Primary productivity rivals that of the nearby forest, but individual plants don't persist—phytoplankters live for days, not centuries. There is no place to store growth except in tiny cells; the biomass of living plant material below a square meter of Puget Sound surface is but a thousandth of that above a square meter of forest floor.

Zooplankters, too, must cope with these facts of scale. Planktonic animals fall into three size classes: the micro-zooplankton, including the protozoa and rotifers; the meso-zooplankton, including the copepods, medusae, ctenophores, chaetognaths, and larvaceans; and the micronekton (animals almost large enough to be true swimming nekton), including the euphausiids, mysids, amphipods, and pteropods. Each of these zooplankters must be the right small size to extract its particular diminutive food from the dilute soup in which they all are suspended. In addition, each must be adapted to the fluctuations in its food supply, which could be bountiful one week and nearly nonexistent the next. These adaptions are reflected in animal reproductive cycles—the more variable the food supply, the more frequent the population's adjustment by means of reproduction. Microzooplankters eat the tiniest plants and multiply every few days. Animals of the micronekton, in contrast, capture prey as large as fish larvae and live for more than a year. These fundamental adaptations are the foundation for the functioning of Puget Sound as a unified ecosystem.

Sink or Swim

Plankton concentrates in and depends upon the score of meters just below the water surface, where life is first generated phytosynthetically. To stay alive all plankters must stay afloat; but plankters are heavier than water, and without compensating mechanisms will sink to a deep, dark death. Any species of plankton existing today must have evolved a means of staying off the bottom at least long enough to propagate itself.

The struggle of plankters to stay afloat has a critical relationship to their sizes. The greater the surface area of an organism, relative to its volume, the greater its friction as it moves through the water, and so the slower it sinks—a "parachute" effect. One way to increase relative surface area is to grow eccentric horns, spines, and wings. Another—the strategy intrinsic to all plankton—is to be small. Indeed, the swimming abilities of plankters have a nearly perfect inverse relationship to their sizes.

Phytoplankters combine their strategies to stay afloat with their strategies to obtain and store nutrition. Small size increases the surface area through which dissolved nutrients may be absorbed from the surrounding water. Odd shapes cause cells to spin and tumble as they sink, stirring the water to bring in nutrients. Some phytoplankters have a large cavity, or vacuole, at the center of each cell, reducing its mass and density relative to its surface. The cell's enzymatic machinery can exchange chemicals between the vacuole and the surrounding water, pumping out heavy ions (magnesium, copper, sulfate) and pumping in lighter ions (nitrogen, potassium, sodium, chlorine). Many microalgae are also buoyed by their intracellular energy stores of fat and oil.

Planktonic plants face a particular dilemma, because the sunlight they need is above, and the nutrients are below. The bottomward trickle of phytoplankton cells and the nutrients they have absorbed causes nutrients to be chronically less abundant near the surface than deeper in the water. A cell cannot float indefinitely at the surface, but must remain in motion lest it completely exhaust the nutrients from the small sphere of water around it. The nonmotile diatoms can do this only by allowing themselves to sink, at an average rate of about one meter per day—slow enough to divide several times before losing sight of the sun, but still sufficient to deplete their populations. Diatoms can only thrive when assisted, both in staying afloat and in finding nutrients, by vigorous stirring, as found in the waters of Puget Sound. In calm waters, therefore, an advantage is conferred on the swimming flagellates, which can regulate the depth at which they grow.

Zooplankters are affected less by sinking, since they have more swimming ability. Large copepods have been observed to cruise at a speed of thirty body lengths (10 centimeters) per second, and in short bursts they may spring 150 body lengths (50 centimeters) in a second. The outstanding feat of swimming, and the most perplexing, is the phenomenon of diel vertical migration. Many groups of zooplankton—some copepods, the chaetognaths, and the micronekton—swim up and down in the water on a 24-hour cycle. Euphausiids and chaetognaths spend the day at depths of 100 to 200 meters, rise to the top 100 meters during the night, then return to the depths before dawn.

Various theories about vertical migration have been proposed: that zooplankters save energy by spending days in cool subsurface water; that they avoid predators by staying in the dark; that they are bothered by bright light; or that they allow plants to grow unmolested by day to harvest a larger crop by night. No experiment or theory, however, has done more than demonstrate that any or all of them are possible.

Sinking and swimming, for both plants and animals, are also tied to reproduction. To maintain their populations in Puget Sound, phytoplankters, besides staying afloat, must avoid being washed by currents out to the Pacific Ocean. Cysts in the sediments persist through periods of arrested growth to reseed the surface waters. Planktonic animals play the enormous odds against survival in the lighted upper layer of the sea—which does not lend itself well to protection of young—by producing hordes of eggs and larvae. The extent of parental care in the zooplankton is the harboring of eggs (or larvae by mysids and amphipods) until they can be released in the surface waters where and when food is available, often under cover of darkness. Many zooplankters may also use their migratory strategies, both daily and seasonally, to avoid the strong surface currents which could carry them far offshore. In contrast, planktonic larvae instinctively seek the surface for a free ride to new territory.

Soil in the Sea

Besides plants and animals, all ecosystems have a third major division, the recycling component, which closes the ecological circuit and reconverts dead material to the nutrients that nourish plants. In the planktonic recycling system, independent of the sea bottom, are bits of organic matter—scraps of uneaten food, feces, molted exoskeletons, and corpses— generated by organisms near the surface and collectively called detritus. Mixed with this debris are inorganic particles washed down from rivers and eroding shorelines. Only about one milligram of particles, less than one percent organic, is suspended in a typical liter of Puget Sound water. Though very dilute, suspended matter resembles terrestrial soil, in that it provides a food source and a physical substrate for certain specialized animals and supports bacteria, which decompose dead tissue and return needed nutrients to the water.

For their irregular appearance, detrital particles are sometimes called "marine snow." The organic particles (or organic coatings on inorganic particles) are an attractive food for some zooplankters, both for mobile animals, which snatch bites as they swim, and for more sedentary microzooplankters, which adhere and feed off them continuously. The sinking flakes accumulate by snowballing, picking up plants, animals, and bacteria as they go. Small-scale current patterns also concentrate particles and hasten their flocculation. The aggregates grow as the organisms within them multiply. When they are disrupted, the new smaller pieces begin the same growth process.

Just as important as particles to recycling in the sea, however, are dissolved chemicals. The sea surface is a broth of both inorganic and organic nutrients, in which the ratio of dissolved to particulate organic matter can exceed a factor of forty. Phytoplankters need some of these organic nutrients, including vitamins. Bacteria interconvert the essential elements of life between particulate, dissolved organic, and dissolved inorganic forms. They decompose detritus into nutrients, and they take up other nutrients into their cells, which become food for animals. Nutrients are also recycled by zooplankton and fish, which excrete highly concentrated organic nitrogen and phosphorus, the former being especially useful to plants. Phytoplankters themselves release (or excrete) some forms of organic carbon, including some vitamins, into the water (perhaps accidentally, perhaps not), and these can contribute further to bacterial growth and recycling. Regenerated nutrients sustain primary productivity when the inorganic nutrients have been depleted by phytoplankton growth.

Patchiness

Oceanographers face a major handicap in studying plankton. For all the vigorous mixing in the sea, the water and its contents are never completely homogenized. Organisms in the sea are patchily distributed; they appear like galaxies in the cosmos or clumps of mushrooms in the forest. The most familiar example of aquatic patchiness is the schooling of fishes. Plankton is found in patches as well, ones more difficult to observe than fish schools. The distribution of plankton is so variable, in fact, that the discrepancy in plankton standing stock between samples taken simultaneously, side by side, is likely to be at least a factor of two.

Patchiness arises from both physical and biological causes. Seawater itself is not uniform. The collision of waters with different origins and different chemistries—surface and deep, inland and offshore, and fresh and salt—creates distinct water blobs or "parcels" that coexist temporarily until they mix together. While parcels persist they offer diverse and contrasting habitats for the plankters they contain. Parcels may be found in any size, from millimeters to kilometers, superimposed on and contained within each other. Persistent wind-generated surface current vortices, called Langmuir cells, concentrate floating matter into visible windrows. Such patchiness can be visible in slight variations of color and texture on the Sound's surface. There is even more variation in the vertical dimension than in the horizontal, since vertical mixing currents are weaker. Detrital particles, themselves tiny patches, form inanimately as they sink, by accumulating in small eddies, and by adhering to the surfaces of underwater air bubbles.

Phytoplankton patches form as colored clouds, with daughter cells created faster than they can be stirred away. Patches of nutrients form around "snowflakes," or where animals excrete, fostering the patchiness of plants. Phytoplankters also form thin horizontal stripes at depths where growth conditions are ideal and sinking is slow. Dinoflagellates, which migrate in and out of the surface daily, are patchy in time as well as in space. Patchiness due to swimming ability, moreover, is the rule among larger zooplankters: in addition to their daily and yearly migrations, they also seem to school as fishes do. These swarms are a more efficient means of exploiting their patchy prey, they are part of reproductive behavior, and they may have some defensive value.

Patchiness is so pervasive that it has become an integral part of the way marine ecosystems function. Repeated experiments have led to the conclusion that if phytoplankton was distributed uniformly throughout the surface water it would be too dilute to support animal life. Zooplankters apparently depend on occasional high concentrations of plants—both in space and time—to find enough food, and will apparently cease feeding when phytoplankton standing stock drops below a certain threshold of abundance. At the same time, paradoxically, other studies have concluded that phytoplankters survive this intense grazing only in the refuge of patches. Like settlers on the plains, they are better defended in a protective circle.

The problem of patchiness once again demonstrates the uncertainty of the scientist. When we analyze seawater, unless we are very careful, we destroy exactly what we seek to observe—the ephemeral, fluid structure that was present. Patchiness, from one point of view, is a nuisance that obstructs efforts to construct a clear picture of the abundance and habits of plankton. But patchiness is not an obstacle to knowing what goes on in the sea—it *is* what goes on in the sea. It deserves study in its own right, since it is as much a part of the plankton's world as size, or sunlight, or the stirring action of the water.

Habitats

Oceanographers divide the sea into zones (Figure 4.3), each containing a different sort of physical or biological action. There are big zones and little zones, and zones within zones. There are, first of all, the pelagic zone and the benthic zone. The pelagic zone is anywhere and everywhere in the open water. The benthic zone is the sea bottom. Plankton, fish, and whales are pelagic, whereas clams, starfish, and other bottom dwellers are benthic.

The zone exposed by the retreating tides is benthic and is called the intertidal zone. This is a unique environment because of its periodic exposure to air and direct sun. Plants and animals that live in this zone face severe problems with heat and dehydration. Below the inter-

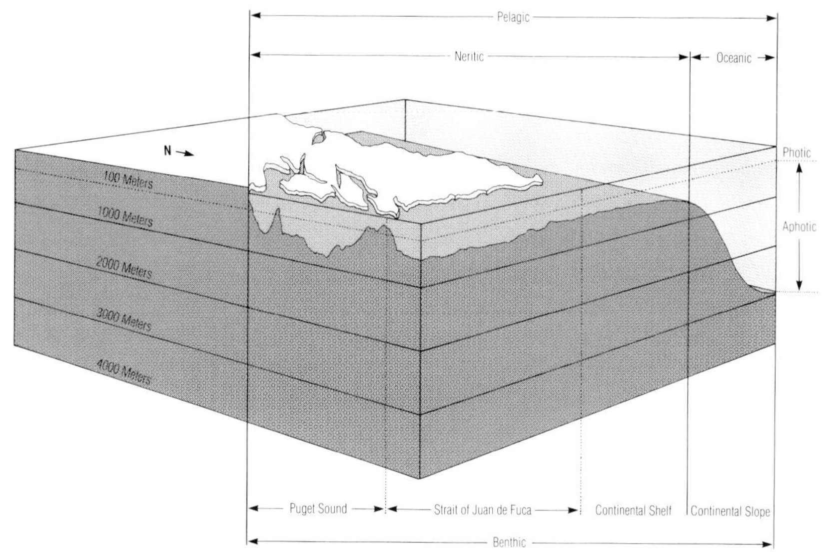

Figure 4.3 This schematic cross-sectional view shows that within the pelagic (open water) domain, Puget Sound, the Strait of Juan de Fuca, and some coastal waters are within the neritic (shallow water) zone. Neritic waters extend offshore to the edge of the underlying continental shelf.

tidal is the littoral zone, where the water is shallow enough that sunlight reaches the bottom and can support seaweeds.

The pelagic zone is divided vertically into the euphotic zone, where there is sufficient light for photosynthesis, and the aphotic zone, where there is not. The boundary between the euphotic and aphotic zones varies in depth; the brightness of the sun and the transparency of the water make it go up and down at different seasons and locations. The pelagic zone is divided horizontally as well. Near shore is the neritic zone, and far from shore is the oceanic zone. The boundary between the neritic and oceanic zones is the edge of the continental shelf, about 50 to 60 kilometers off the Olympic coast, where the water depth does not exceed 200 meters. Thus, all of Puget Sound is within the neritic zone.

The distinction between the neritic and oceanic zones is made because the nature of planktonic life changes in deeper water. The water over much of the shelf is shallow enough that, especially during a storm, nutrients and resting cysts can be stirred from the bottom into the euphotic zone to stimulate phytoplankton growth.

Puget Sound is a special kind of zone within the neritic zone; it is an estuary. Grays Harbor, San Francisco Bay, and Chesapeake Bay are other familiar estuaries, all characterized as semi-enclosed areas where fresh water meets salt water. Puget Sound is also a fjord, an inlet created by a glacier. Fjords are deep, narrow, and steep-sided, and often have shallow plugs called sills left by the retreating glacier. Puget Sound has sills at several locations: at Admiralty Inlet, between the San

Juan Islands, at Deception Pass, at the mouth of Hood Canal, and at The Narrows (page 46).

Biologists divide the world into another set of zones, based on biological boundaries. These maps may coincide with geographical features, but the real zone boundaries are abstract and can't be drawn on paper. An ecosystem, for example, may be defined as a physical space and the living things that live in it, with the hypothetical boundary of an ecosystem drawn where creatures are not affected by either the organisms or the environment beyond its borders.

Puget Sound, neatly defined by surrounding land and narrow connections to the Pacific Ocean and the Strait of Georgia, appears to stand as an ecosystem unto itself. But the definition is not a perfect fit. Puget Sound and the Strait of Juan de Fuca are not biologically independent of either the Pacific Ocean or the Strait of Georgia nor of the lands bordering all of them. The water and organisms exchanged among these water bodies intertwine all of their ecological fates. Symbolizing this biological unity is the Pacific salmon, which makes all of the coastal and northernmost North Pacific its home. Furthermore, Puget Sound is not uniform within its boundaries—plankton behaves very differently in the shallow fingers of the southern Sound, for instance, than in the open Strait of Juan de Fuca. Nevertheless, the physical properties controlling plankton, and the biological interactions that result, are constant enough within the Sound and different enough from the open Pacific to merit separate consideration.

Within an ecosystem such as Puget Sound, the fates of plants and animals are subject to two broad biological processes, competition and predation. Each organism must contend with other, similar organisms for common needs—sunlight, nutrients, food, etc.—at the same time that animals are trying to eat them.

Superficially, the world of the plankton might seem to be a single arena in which all the species compete. If this were so, we might expect a steady disappearance of species, a winnowing of the weak from the strong. The "Competitive Exclusion Principle" states that given enough time one superior species should exclude all others with which it competes. In Puget Sound, however, as in the sea at large, different plankton species wax and wane. Although one species may dominate almost completely under some circumstances, it does not persist; other species make comebacks. Dominance in the plankton is the exception, and diversity is the rule. The Competitive Exclusion Principle is clearly inadequate to describe Puget Sound.

G. Evelyn Hutchinson, one of the grand old men of aquatic biology, called this failure "The Paradox of the Plankton." He suggested that the principle never gets sufficient time to take hold. Before a dominant species has a chance to exclude its competitors, conditions change and a new species begins to take over. Such variability is built into natural systems, on time scales of hours, weeks, and even years, and leaves the door open for multiple species to maintain their footholds.

Conditions in the plankton vary in space as well as in time, and that patchiness permits competing species to dwell contemporaneously in slightly different habitats. Furthermore, under most conditions in the sea, competing plankters may be sufficiently dilute that they do not even feel each other's presence, living within tiny spheres of influence which may never overlap. Some plankters adopt behavioral traits—migration patterns, differing nutrient and food requirements, and organism size—which partition the surface waters into delicately carved niches where species can avoid competition. Superimposed on all of these intrinsic influences are the extrinsic effects of predators. Animals can be highly selective of their food, and so can skew ecological competition toward organisms with the best defenses.

Food Chains

Various tangible objects have been used as models for visualizing the ecological relationships by which higher organisms obtain nutrition from lower ones. None of them can fully portray the complex and dynamic nature of an actual ecosystem, any more than a pencil sketch can portray a human being. A useful portrait, however, conveys the most information with the fewest lines.

The simplest ecological model is the food chain, by which each organism is linked to its sources of nutrition and to the organisms that feed on it. Food chains are linear and one-dimensional, with each organism assigned to a numbered trophic level. At the first trophic level are the plants, which as autotrophs make their own food. Heterotrophs, which must obtain food from other organisms, are assigned to the second trophic level of the herbivores and to the third and higher levels of the carnivores.

More complex and realistic is the food web, in which animals do not fall neatly into trophic levels. Omnivores (which feed on both plants and animals), animals that change their feeding habits as they mature or as food abundance changes, and predators that feed on other animals ostensibly at the same trophic level all form a second dimension of cross-links between many coexisting and changing food chains. An animal may obtain its food through several different routes simultaneously, each perhaps having a different number of trophic links. Despite this increased complexity, however, food webs are still very crude representations of systems which defy even the power of a computer to portray them.

Food chains and webs at the sea surface are unlike any other on earth. The pelagic food chain is rigidly built according to size, with pelagic animals almost always larger than their food. In forests, the largest organisms are the massive long-lived trees. In grassland and tundra ecosystems, the largest organisms are long-lived, highly mobile herbivores such as bison and caribou. In Puget Sound, however, the largest organisms are carnivores. Only killer whales—which can gang up and steal the tongue of the larger minke whale—ever use the strategy of the wolf or the lion, and hunt in packs for large prey.

Paradoxically, the trophic relationships of highly fertile Puget Sound most closely resemble those of a desert or dry prairie. In both, most plants are relatively small and patchy, their growth is intermittent, and their abundance changes at the whim of the physical elements. Many planktonic and desert animals hide during the daytime and feed at night. As in Puget Sound, herbivores of the desert (such as rodents and insects) are small and opportunistic, carnivores (such as birds) are large, fierce, and mobile, and very little biomass accumulates either in living organisms or in organic detritus.

To grasp the essential information about how animal populations in Puget Sound are controlled by both the abundance of food and the path by which it is obtained, it is necessary to begin with the simplest model, the food chain. This approach will sacrifice a great deal of interesting detail. Eventually some of the most important features of Puget Sound's pelagic ecosystem will only be explainable by adopting the more complex picture of a food web.

CHAPTER FIVE

The Green Machine—Phytoplankton

Nihil vilior algae (Nothing is more worthless than algae).
Virgil

Puget Sound is a solar-powered factory in which animals are assembled from raw materials of water and the carbon, nitrogen, phosphorus, and other vital chemicals dissolved in it. The analogy is unfair in some respects, since a single cell, with its ability to replicate, is more wondrously complex than the largest factory. Much can be learned, nevertheless, by examining organisms and ecosystems as though they were mechanical.

Puget Sound's pelagic food chain, which culminates in fishes, birds, mammals, and humans, begins with the phytoplankton. This community of tiny plants is a biological antenna deployed over the entire surface of the Sound, gathering the solar energy needed for the production of animals. The phytoplankton is also a rechargeable storage battery, which photosynthetically sequesters sunlight in the chemical form of combined carbon. Solar energy stored in biomass can be called trophic energy. Thus carbon becomes the fuel and, along with the nitrogen in protein, the structural material for fabricating the entire living ecosystem.

To comprehend the Puget Sound pelagic food chain as a biological machine, and to trace the path of biomass—"food"—and the trophic energy it contains as they are transmitted through each trophic level, we must examine some of the important environmental and biological constraints which regulate that transmission. The starting point is the interface between the phytoplankton and its surroundings. Plants are directly tied to their environment, whereas animals are, to some extent, segregated from it, buffered by their position on the food chain.

The Physical Setting

The understanding of primary production in Puget Sound depends as much on physics as on biology. Like terrestrial plants, planktonic plants need only enough light and enough nutrients and they will grow. Unlike the land, however, Puget Sound is not at rest, and water motion affects the availability of both light and nutrients. Phytoplankton growth in the Sound is regulated by the interaction between sun, precipitation, and forces that set water moving, including wind. In short, as on land, plant production in water depends on the weather.

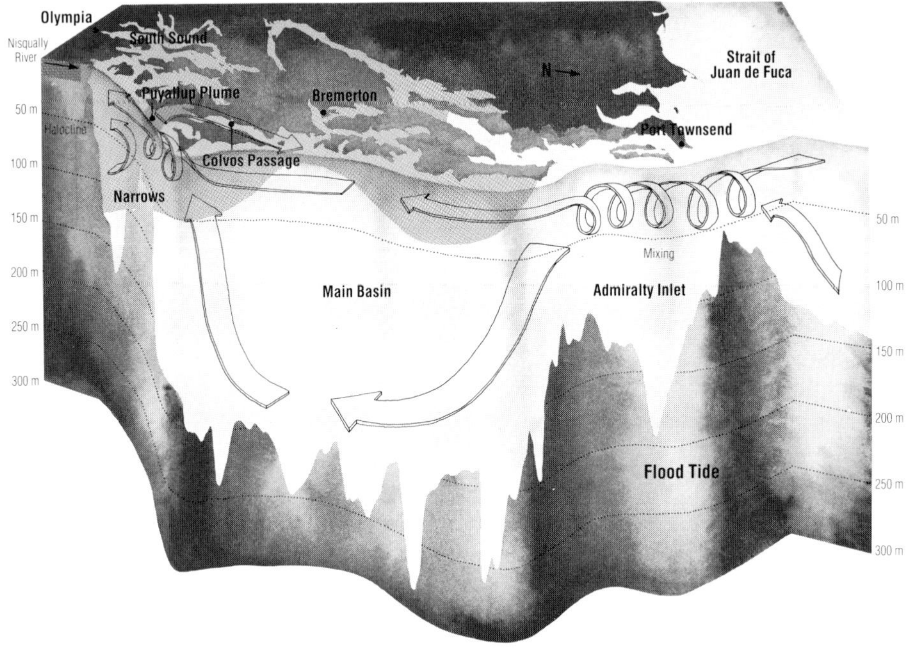

Figure 5.1A Currents in the main basin—flood tide. Intruding flood tides, with water volume over a hundred times greater than river runoff, produce landward currents in the main basin. The flow is slightly altered around Vashon Island: currents are seaward in Colvos Passage at all tidal stages. Mixing over sills occurs on both flood and ebb tides.

Weather is notoriously unpredictable, on time scales of both hours to days, and years to centuries. Forecasting plankton growth is even more speculative than the weather forecasting on which it depends, because the historical records of plankton have been sporadic, arbitrary, unreliable, and incomplete. They have the quality of weather data from a century ago. Until better data are collected, there will continue to be more mysteries than solutions, more clues than evidence, and more suspicion than proof.

Mixing versus Stratification

At the heart of the relationship of planktonic plants to their liquid milieu is the process of mixing. Water churning through narrow passages in Puget Sound, homogenized like the contents of a kitchen blender, demonstrates how intense the mixing process can be. The pattern of primary productivity in the Sound closely matches the pattern of water mixing. Specifically, it is vertical mixing, between surface and underlying waters, that is of most importance to phytoplankton production.

Vertical mixing can have a negative effect on productivity—stronger and deeper mixing reduces the exposure of phytoplankton to bright surface sunlight, and slows the rate of photosynthesis. Vertical mixing

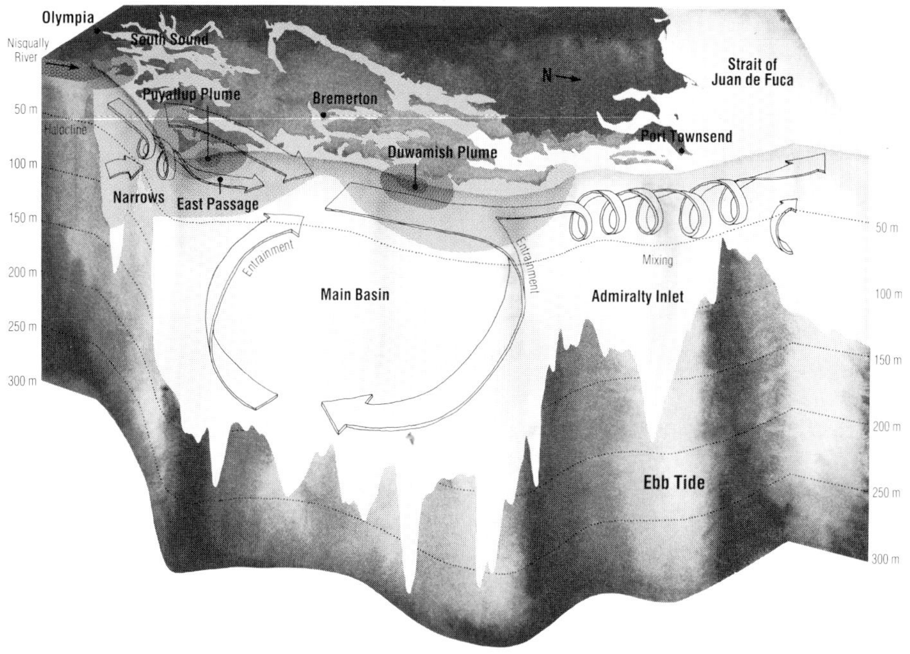

Figure 5.1B Currents in the main basin—ebb tide. In deep water, there is weak landward flow. Surface water flows seaward, entraining bottom water into its lower layers. Roughly two-thirds of the surface flow reaching Admiralty Inlet returns to the main basin as it mixes with deeper inflow. Low salinity river plumes (dark shading) are evident off the Nisqually, Puyallup, and Duwamish rivers. Currents in East Passage remain predominantly landward at all tidal stages. (After Barnes and Ebbesmeyer, 1980)

can also be beneficial because it brings nutrients from deep water, where they are plentiful, toward the surface, where they are most needed and yet likely to be scarce. The role of vertical mixing, then, is twofold: too much mixing depresses photosynthesis and disperses phytoplankton populations, but not enough mixing leads to nutrient exhaustion. A little bit of mixing is just right, and this is what makes Puget Sound such a productive ecosystem.

The agent of both horizontal and vertical mixing is moving water. When seawater stands still, the only transport processes—molecular diffusion of dissolved chemicals and sinking of suspended matter—are relatively slow. Any physical force that sets water into motion, particularly turbulent motion, greatly increases the rate of mixing.

Currents, including tidal currents, are the dominant water motions in Puget Sound. The pattern of net circulation, averaged over at least several tidal cycles, is typical of an estuary (Figure 5.1). River runoff drives fresh water seaward at the surface, and pulls along (entrains) several times its own volume of the underlying salt water. Deeper water, as a consequence, flows landward from the Strait of Juan de Fuca to replace the salt water entrained in the surface outflow. The net flow changes from seaward to landward at a depth of roughly 50 meters. The inexorable skimming off and replacement from below of the water in this transition zone produces a gradual upward mixing of nutrients.

At any given moment, the mean current condition is hidden amidst the much stronger twice-daily rhythm of tidal currents, which can completely reverse the estuarine outflow. Surface water flows two steps seaward on the ebb tide (Figure 5.1), then one step back on the flood tide. Likewise, deep inflow is retarded by the ebb and reinforced by the flood. This seesawing of waters from the rivers and the sea produces both horizontal and vertical mixing where the two meet.

Tidal currents and the mixing they cause vary in strength with the tidal amplitude. Each month there are two periods of "spring" tides (periods of especially high and low tides) associated with the new and full moons, and two "neap" tides of low amplitude during the quarter moons. The spring tides during the months of April, May, and June are of particularly large amplitude, and thus cause bimonthly periods of especially vigorous mixing when the growth season is just beginning.

Vertical mixing is strongest where the ebb and flood rush over shallow shoals. The forcing of currents over sills and through narrow constrictions propels deep water upwards, causing strong mixing often visible as surface turbulence. The sill at Admiralty Inlet, in fact, obstructs the continuous inflow of deep water, which pulses intermittently instead. Although this process occurs at only a few places, these are the dominant areas of mixing for the entire Sound.

Winds also cause mixing. Puget Sound is more sheltered from wind action than coastal and open ocean waters, but in some cases—the storm that destroyed the bridge over Hood Canal in February, 1979 is an outstanding example—its effects on the surface can be severe. The wind acts by raising waves, which stir the surface layer. The wind also generates currents in the direction it blows, and since the wind direction may not be the same as that of the underlying surface current, it can alter the existing water flow and the resultant mixing. The effects of winds are most pronounced when their direction is parallel to the long axis of a body of water—that is, when they have a long fetch over which to act. The 1979 storm was devastating because the wind blew directly down the long axis of Hood Canal, in opposition to an incoming tide, and built the water up to an extreme crest. In most of Puget Sound, northerly and southerly winds have the greatest effects.

Given enough light and nutrients, rapid phytoplankton growth can commence at Puget Sound's surface simply due to the absence of strong mixing. Mixing forces are pervasive, however, and little growth usually occurs unless mixing is counteracted by some stabilizing force that permits phytoplankters to bask undisturbed in surface sunshine. Surface water stability—its resistance to vertical mixing—is the result of stratification, the presence of a layer of low-density water floating atop denser water. The boundary between the two layers (the pycnocline) is a physical barrier, which limits the depth to which surface water and phytoplankton are carried by all but the strongest mixing forces. The greater the density difference between the two layers, the stronger the stratification and the greater the stability.

Puget Sound and other estuaries are highly productive partly because of the stabilizing effects of brackish river runoff, which flows seaward atop denser saline water. The boundary between the two, at about 50 meters in the main basin, is called the halocline (Figure 5.1). Reinforcing this saline stratification, occasionally even dominating it, can be shallower thermal stratification due to less-dense, sun-warmed surface water. Stratification is most important near the surface (in the upper 20 to 25 meters), where most production takes place. Stratification and the accompanying stability persist until strong mixing forces intervene to disrupt them.

For sustained phytoplankton growth, there must be a balance between mixing and stratification; that is, mostly between fresh water and the various forces that mix it with salt water. Without sufficient stratification, phytoplankton growth is suppressed, but if stratification is too persistent, nutrients become depleted and productivity diminishes.

Puget Sound experiences a highly productive balance because mixing and stratification are patchy in space and time. This physical patchiness results in biological patchiness as well. Lush phytoplankton oases are separated by unproductive gulfs of space and time. Most phytoplankton growth in the Sound appears as dynamic outbursts of productivity called blooms, interspersed with areas or times of marginal growth. An intense bloom usually occurs in the spring, and perhaps another in the fall. In the winter, plants are sparse, and in the summer there is not so much steady growth as a series of intermittent blooms, as short-lived organisms come and go. The best illustration of the linkage of biological to physical patchiness is provided by studying the blooms in Puget Sound's main basin in the spring.

The Main Basin

The main basin of Puget Sound (Figure 5.2) is the deep, wide basin that extends about six kilometers from Seattle on the east to Bainbridge Island on the west, and about 66 kilometers from Admiralty Inlet (actually a strait) on the north to The Narrows on the south. There is a sill at a depth of 60 meters below Admiralty Inlet, another at 40 meters below The Narrows, and the maximum depth of the basin exceeds 250 meters off Shilshole Bay. The main basin serves as a prototype for all of the physical and biological processes that govern primary productivity throughout the Sound. Other regions of the Sound demonstrate nearly

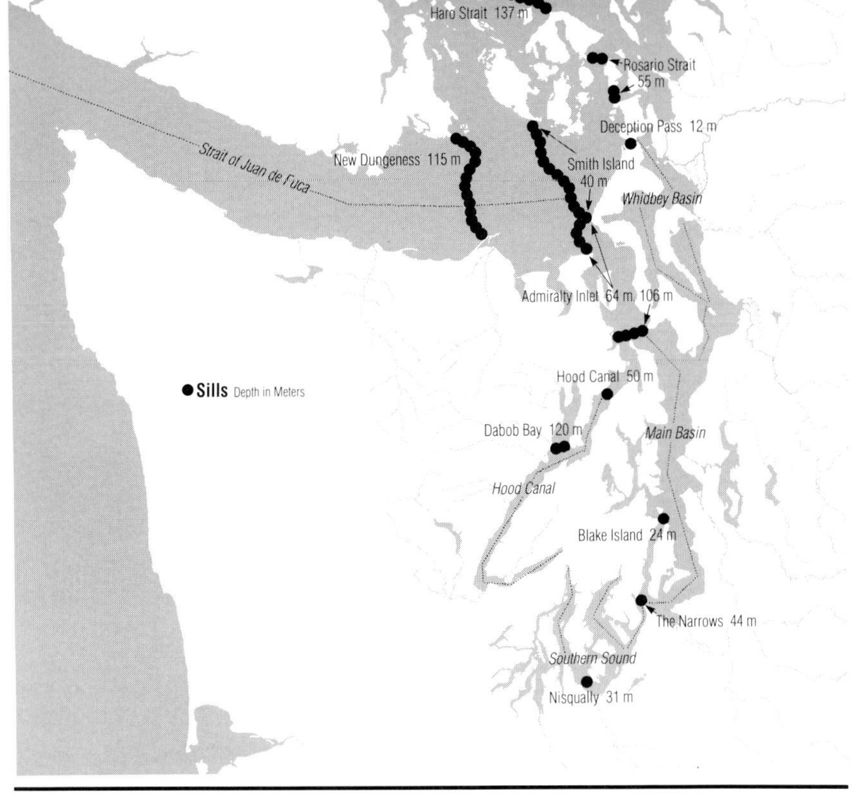

Figure 5.2 Basins and sills of Puget Sound. Shallow sills separate Puget Sound from the Straits of Georgia and Juan de Fuca, and separate the main basin from Hood Canal and the southern Sound. (After Barnes and Ebbesmeyer, 1980)

every possible permutation of the physical processes driving the ecosystem.

The first process is surface stratification, brought about by fresh water. Two-thirds of the fresh water reaching all of Puget Sound enters the main basin from the Skagit and Snohomish Rivers via the Whidbey basin, with smaller amounts contributed by Lake Washington and the Duwamish and Puyallup Rivers. Additional fresh water, chiefly from the Nisqually River, enters through the southern Sound via The Narrows. The resulting brackish layer flows generally northward along the surface. The rate at which rivers empty water into the Sound is determined by both rainfall and snowmelt, and varies considerably with the seasons. Runoff is lowest in late summer, while the peak of rainfall is in winter and the snowmelt peak reaches the Sound in May and June.

Thus one condition necessary for phytoplankton growth—stratification—is present in the main basin through the winter and spring. Growth begins when this stratification coincides with the spring increase in the amount of sunlight (Figure 5.3). Unlike runoff, solar input is lowest in winter and highest in summer. Both sunlight and stratification are near their annual maxima during the spring, making it the prime season for phytoplankton growth.

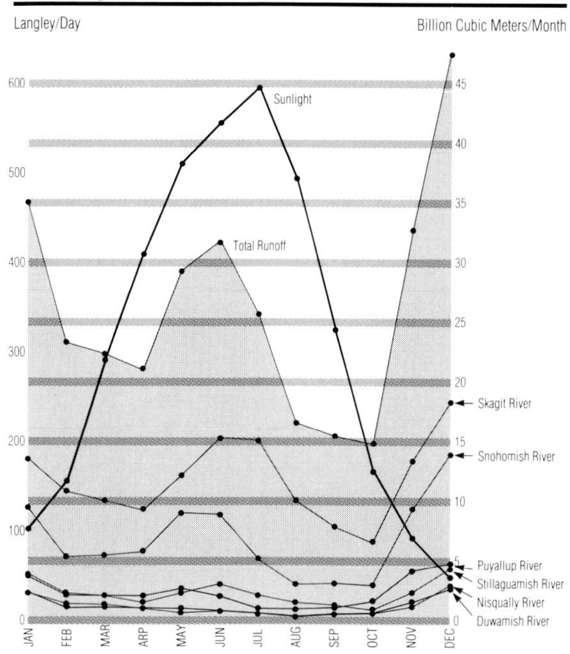

Figure 5.3 The growing season in Puget Sound. Increased sunlight caused by longer days, clearer skies, and a higher sun angle coincides with stratification at the surface, making May to July the peak growing season in the main basin. Freshwater runoff accounts for stratification, and is shown here as outflow from individual rivers and as total input to the main basin. (After Dexter et al., 1981; and Ebbesmeyer and Helseth, 1977)

There are further complexities to the regulation of primary productivity in the main basin, however. The onset of phytoplankton growth occurs later here than in other estuaries at the same latitude, or even in other regions of the Sound; yet on an annual basis the main basin is more productive than these other areas. Furthermore, the growth occurs as blooms, which like sunshine and runoff are discrete events whose timing and intensity differs considerably between years and may have little resemblance to a long-term average condition. To understand these irregularities requires a closer examination of the unique physical environment of the main basin.

Sills and Residence Time

The main basin of Puget Sound is bathymetrically embraced, as though by a pair of bookends, by shallow sills at its northern and southern ends. The sills alter the normal pattern of estuarine circulation by causing mixing and by restricting the exchange of water with adjacent basins, and these alterations contribute to the singular patchiness and productivity of the main basin.

Not all of the surface outflow in the main basin comes directly from rivers. Fresh water from the southern Sound must pass through The Narrows on its seaward course. From The Narrows outflow traverses Colvos Passage and encounters another sill near Blake Island before entering the main basin. Along this route water from a depth of 120 meters (80 meters below the sill depth) is completely blended with surface water, producing roughly a quarter of the main basin's outflow. This disruption of stratification dilutes whatever phytoplankton grows in the southern Sound, and suppresses primary productivity at The Narrows throughout the year. It also accounts for the delay of blooms until April and May in the main basin, compared to estuaries with similar runoff and sunlight characteristics, such as Long Island Sound, where blooms begin in March.

In the open main basin, mixing forces are weaker, and phytoplankton can grow using the rich nutrients pumped to the surface at The Narrows. Once sunlight is sufficient to compensate for the weak stability, these nutrients sustain primary productivity at a high level through the summer. The annual production of the main basin exceeds that of Long Island Sound by a third or more. This, then, is the secret of the main basin's high productivity: vigorous mixing is patchy, restricted to a small upstream area, and growth proceeds undisrupted in the open basin.

The surface water in the open main basin is also quite patchy, coming as it does from two different sorts of sources. Highly stratified river runoff flows seaward side by side with less stratified but nutrient-rich Narrows water. These discrete waters appear as "stripes," greener as they drift north, which are generated twice a day by the ebbing tide. Depending on the strength of tidal and other currents, the stripes are about nine kilometers long, and can be visible from a ship or an airplane (Figure 5.4). The two different source waters represent the two extremes of physical conditions for phytoplankton growth, mixed waters and stratified waters. Seemingly the most favorable spots for phytoplankton growth would be where these streams make contact, where both nutrients and stratification are in close proximity. Perhaps such contacts could even be traced by locating an optimal salinity that marks their boundary. Such speculations aside, it is clear that "stripes" make the surface of the main basin a highly heterogeneous environment.

To understand patchiness in time, some further details of the circulation in the main basin are needed. Stripes have a finite life span. Oscillating with the tides, surface water traverses the length of the main basin and reaches Admiralty Inlet in roughly six days—faster when runoff is profuse, slower when held back by northerly winds. There the stripes encounter another sill, which once again homogenizes surface and deep waters and disperses surface patchiness, stratification, and blooms. This disruption further alters the classical picture of estuarine circulation. A major fraction of surface outflow is believed to be diverted back into the Sound, rather than to the Strait of Juan de Fuca. About two-thirds of the deep water entering the main basin, in fact, is thought to be main basin surface water caught in the deep inflow during mixing at Admiralty Inlet, rather than water from the Strait.

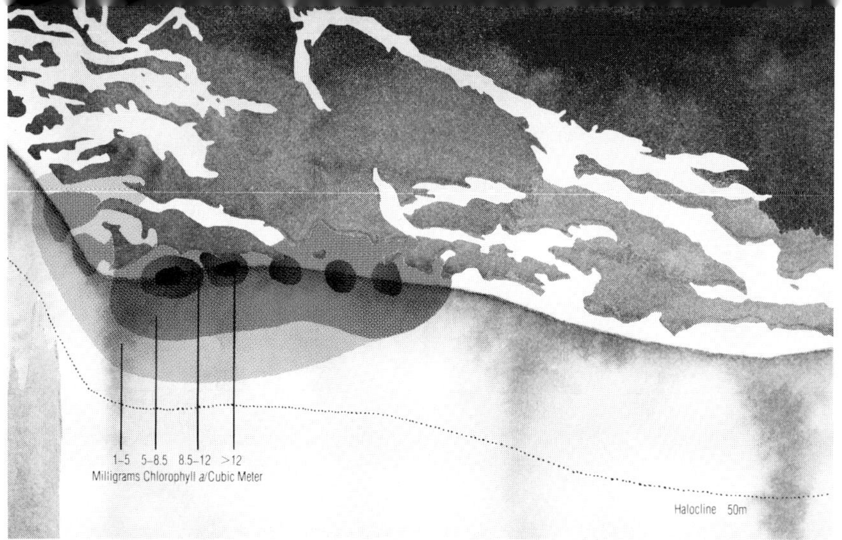

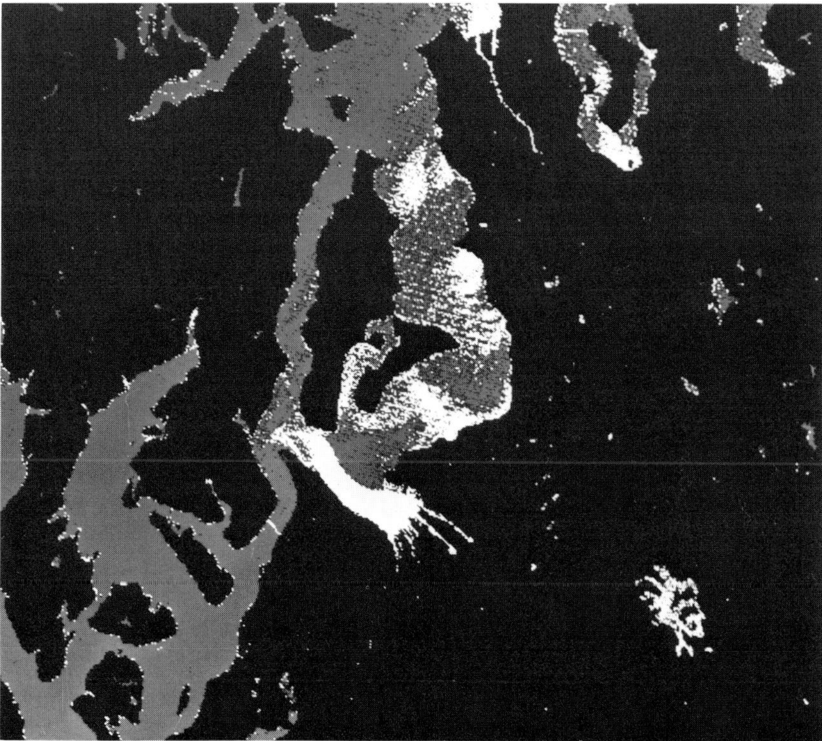

Figure 5.4 Patchiness in the main basin. Top: Tidal striping of surface chlorophyll patches as derived from continuous fluorometric measurements on May 15, 1969 (After Munson, 1970).

Bottom: Computer-enhanced satellite (LANDSAT) photograph of the main basin on June 13, 1974, showing surface tidal striping off the Puyallup River. The photo shows light of wavelengths of 600–700 nanometers (billionths of a meter) and detects both chlorophyll and nonliving suspended sediment. (Photo courtesy Smyth Associates, Inc.)

49

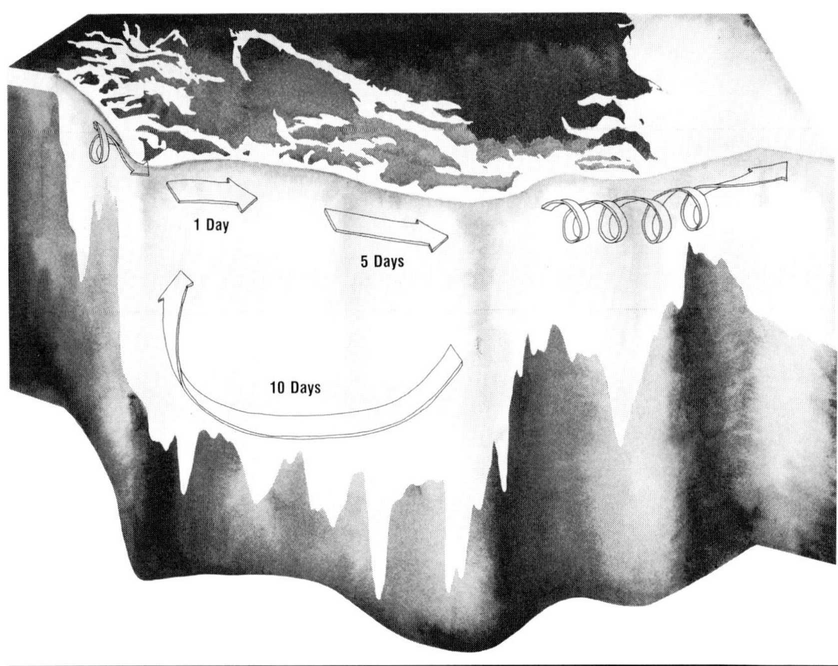

Figure 5.5 Entrainment of surface water by intrusions from the Strait of Juan de Fuca and of deep water by upwelling at The Narrows produces a semi-closed loop in the average circulation pattern of the main basin. Water and plankton average one trip through the deep portion of the loop before exiting at Admiralty Inlet (After Ebbesmeyer and Helseth, 1977).

Thus the mixing at Admiralty Inlet partially short-circuits the two-layer estuarine circulation pattern. Water that enters the Sound at a river mouth might receive an infusion of Pacific salt water at Admiralty Inlet, be carried back through the cold, dark depths of the main basin to The Narrows, then be spurted back to the surface, perhaps repeating the cycle several times before finally exiting seaward (Figure 5.5). The travel time through the deep basin is longer—ten to twenty days—than at the surface. This semi-continuous loop pattern of circulation resembles a conveyor belt, carrying water, salt, nutrients, and phytoplankton back and forth between The Narrows and Admiralty Inlet, between the surface and the depths.

The retention and recycling of water within the main basin have several biological consequences. The salinity and nutrient content of deep water are lower than they would be with more inflow from the Strait of Juan de Fuca, reducing stratification, nutrient content, and productivity of surface waters past the spring and into the summer. At the same time, because phytoplankton is carried into deep water, there is more total chlorophyll below the euphotic zone than within it. This reservoir tends to offset the reduction in productivity by retaining some of the phytoplankton generated, and by seeding water upwelled at The

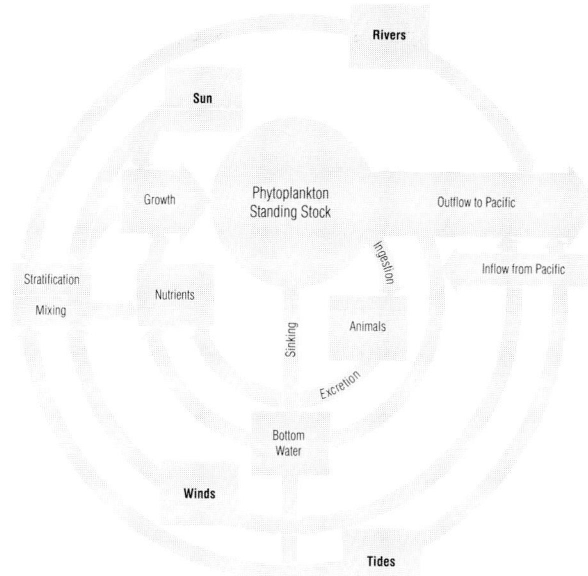

Figure 5.6 Physical forces produce both positive and negative effects on phytoplankton growth, and blooms occur when the forces are balanced. Phytoplankton standing stock is increased by photosynthesis and nutrient uptake, which are regulated by sunlight, stratification, and mixing. Standing stock is decreased by animal consumption, sinking, and flushing by winds, tides, and runoff.

Narrows with an increased stock of cells that survived suspended animation deep in the water. Such a cycle, enhanced by the vertical migration of dinoflagellates, has been observed in another, shallower estuary, Chesapeake Bay. The greatest biological consequence, however, is the effect of the liquid conveyor belt on the timing of blooms.

Residence time (also called retention time or flushing time) is a third major factor, in addition to sunlight and stratification, that governs phytoplankton blooms in the main basin. Surface residence time is the duration of the transit along the surface portion of the liquid conveyor belt, from a river mouth or from The Narrows, where surface water is generated, to Admiralty Inlet, where it is dispersed. While the interplay between sunlight and stratification controls the rate of photosynthesis, the residence time controls the standing stock, which accumulates as a product of photosynthesis. Residence time effects have also been observed in estuaries in Maine, off New York, in San Francisco Bay, and in India, as well as in Puget Sound marinas.

The effects of residence time on phytoplankton growth resemble those of mixing. As with mixing, intermediate residence, rather than long or short, is optimal. Phytoplankters need several days of uninterrupted growth to build their standing stock to bloom proportions, especially when beginning from the sparse populations of a Puget Sound winter. A short residence time means a short exposure to sunlight, and a low resulting standing stock. When both sunlight and stability are favorable, very long residence time, in contrast, leads to a slow rate of nutrient input at The Narrows, nutrient depletion, and inhibition of phytoplankton growth.

There are no measurements of its variability, but the residence time of main basin surface water is governed, in complex fashion, by the same agents of water motion that cause mixing: river flow, tides, and winds. Productivity in the main basin might be even higher if runoff and residence time effects did not often tend to counteract each other. Heavy runoff, which reduces mixing and creates a stable layer for phytoplankton growth, also flushes nascent blooms rapidly out of the main basin. Weak runoff and longer residence coincide with weaker stability. Tidal effects on residence times parallel those on mixing—the stronger currents during the bimonthly spring tides increase the rates of both flushing and mixing, and so reduce the potential for blooms. Finally, all winds increase surface mixing, but the southerly winds of winter and of storms accelerate surface outflow and reduce residence time in the main basin, while the northerly fair weather winds that accompany sunshine also benefit productivity by lengthening residence time.

All of these influences are highly variable in time, and produce the extreme temporal patchiness of phytoplankton growth and abundance in the main basin characterized as blooms. The effects of all these physical forces on primary productivity are summarized graphically in Figure 5.6. The occurrence of a bloom depends on the coincidence of sunlight with some or all of the physical forces that favor intermediate stability and residence time: moderate runoff, neap tides, and light northerly winds. An abrupt alteration in one or more of these beneficent influences could prevent a bloom, or cause a nascent bloom to collapse.

The prime season for blooms is in the spring, because these physical influences do not occur completely randomly or independently of each other, but tend to co-occur. Increasing springtime sunshine is associated with both increased runoff and northerly winds. The balance of these influences is delicate enough that the bimonthly advent of the neap tides can trigger a bloom most often in late April or early May. The variability of blooms in a given year and in different years, and the sensitivity of the main basin ecosystem to its various physical influences, are illustrated by case studies of individual years.

The Spring Bloom
Blooms have a strong element of randomness to their appearances. Although the timing of the spring bloom depends on some quite predictable events, such as neap tides, and on fairly reliable increases in sunlight, runoff, and northerly winds, these seasonal trends only specify an envelope of probabilities. The exact timing of a clear spell, or the onset, magnitude, and duration of a consequent bloom, are largely matters of chance. As important to understanding primary production in the main basin of Puget Sound as a listing of the driving forces, is a

knowledge of their variability. One day does not necessarily look like the next, one year does not look like another, nor is any of them likely to match the long-term average condition. Furthermore, the ability to comprehend this variability is confounded by the practical limitations on sampling.

Figure 5.7 shows the smallest scale of time variability: sunshine, runoff, stratification, tides, and phytoplankton at a single location can change markedly within a few days. Blooms—outbursts of growth lasting ten days or less—appear at intervals during the study period, April through June. They appear at different times in each of the years illustrated, reflecting the different timing of tides and weather.

The brief period of observation during 1969 appears to have been ideal for bloom formation. Bright sunshine, moderate runoff and stability, and gentle northerly winds all coincided with a neap tide. Although the conditions (or the data collection) did not continue long enough to furnish a high biomass, the primary productivity (nearly ten grams of carbon per square meter on May 10) approached the highest ever measured in marine phytoplankton. The bloom appears to have been terminated by an abrupt increase in runoff, which although increasing the stability also flushed the surface waters rapidly seaward.

The opposite extreme, a very unproductive spring, occurred in 1975. Low sunlight and high runoff that year dictated small blooms, which nevertheless corresponded fairly closely with neap tides and high stability.

Perhaps 1966 was a more representative year, and one in which the interaction of the various bloom-forming forces is well illustrated. Although the first and most extensive bloom of that season also appears to have been flushed away, the ensuing four phytoplankton peaks correspond closely to moderate peaks in stability. Runoff was generally low, sunlight generally high (especially in early June), and tidal ranges low (if not exactly neap) at the time of each bloom.

Each of these examples differs significantly from the composite seasonal picture obtained by summing a decade of data. This is the statistical envelope of bloom probability; it is an average year, but not a typical year, because sun and phytoplankton arrive in discrete events, not in smooth continuous gradations. Nevertheless, the graph presents useful information concerning the entire year's primary production. In the average year, 465 grams of carbon are fixed by phytoplankton over each square meter of the main basin's surface. The most productive period is likely to be during May, and 86 percent of the year's production can be expected from April through August. The midpoint of the growing season falls around the summer solstice, and the average stratification is symmetrical about that date. Yet there is consistently reduced production around that date, and again later in the season. Although

Figure 5.7 Temporal variability of phytoplankton and environmental forces in the main basin. Measured sunlight, estimated total Puget Sound river runoff, predicted maximum tidal range, measured stability (the density increase from the surface to 25 meters depth), and the resulting measured phytoplankton production and biomass are shown as they vary on three time scales. The first is the 90-day spring bloom period in three representative years. The second is the mean annual pattern obtained by averaging data by month for the decade 1966–1975. The third shows differences between years in the same decade, from either monthly or annual averages. (After Campbell et al., 1977; Coomes et al., in press; Ebbesmeyer and Helseth, 1977; Harris, 1981; METRO, unpublished; Munson, 1970, U.S. Environmental Data Service; and Winter et al., 1975).

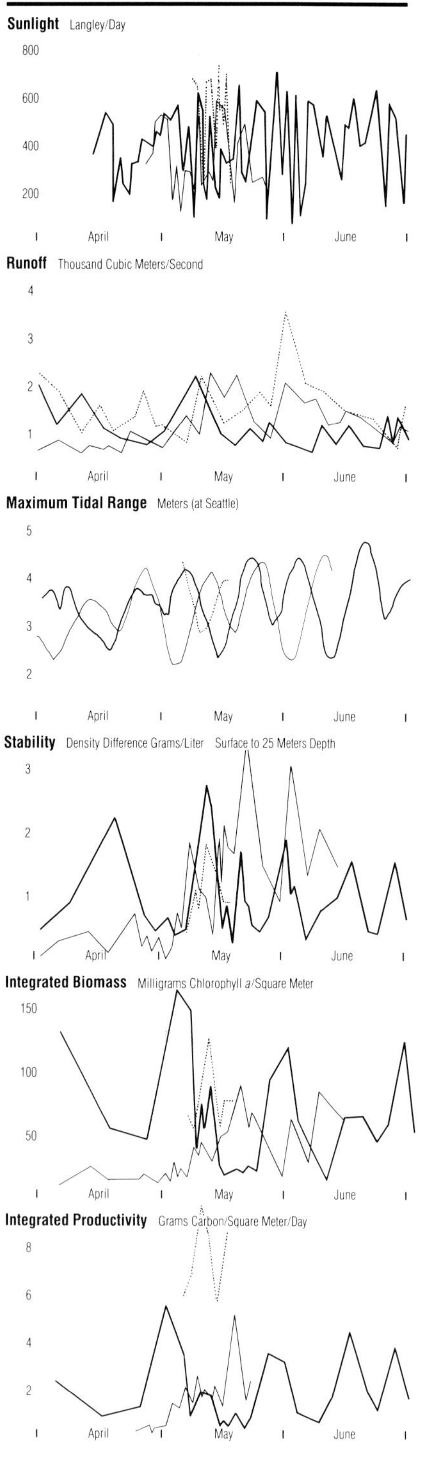

Monthly Changes

Sunlight Langley/Day

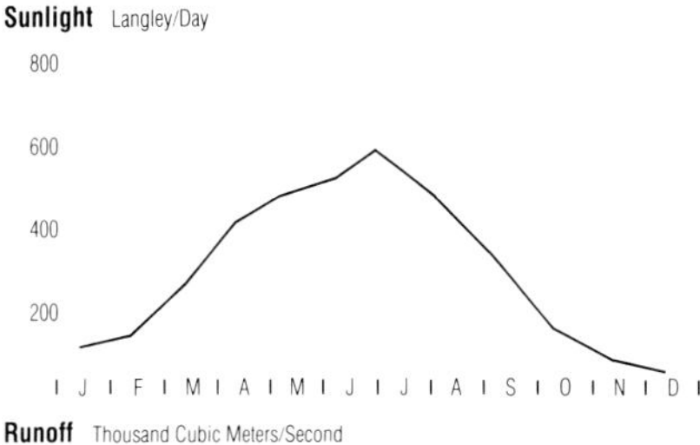

Runoff Thousand Cubic Meters/Second

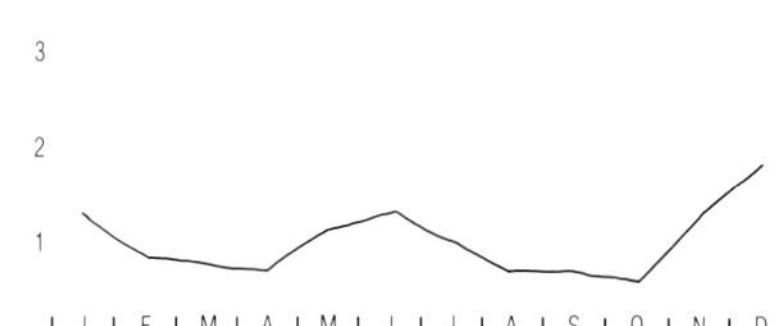

Maximum Tidal Range Meters (at Seattle)

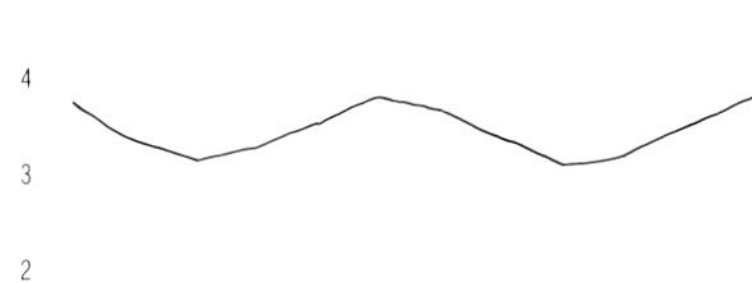

Stability Density Difference Grams/Liter Surface to 25 Meters Depth

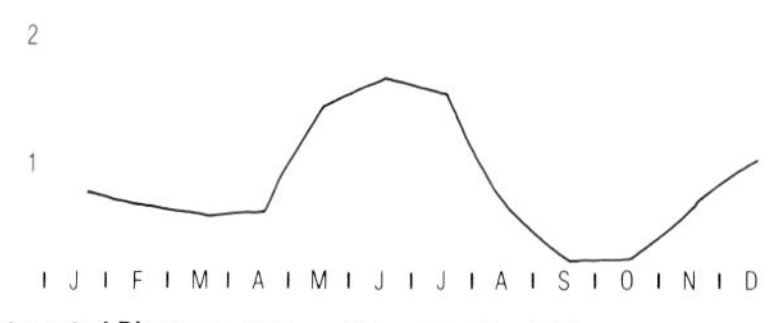

Integrated Biomass Milligrams Chlorophyll *a*/Square Meter

Integrated Productivity Grams Carbon/Square Meter/Day

Yearly Changes

Sunlight Langley/Day (By Month)

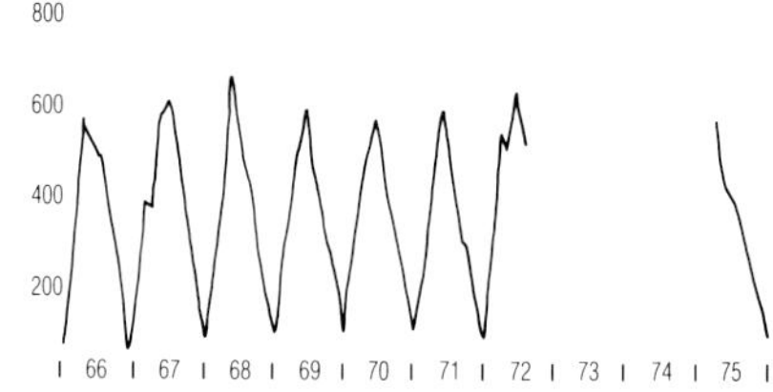

Runoff Thousand Cubic Meters/Second (By Month)

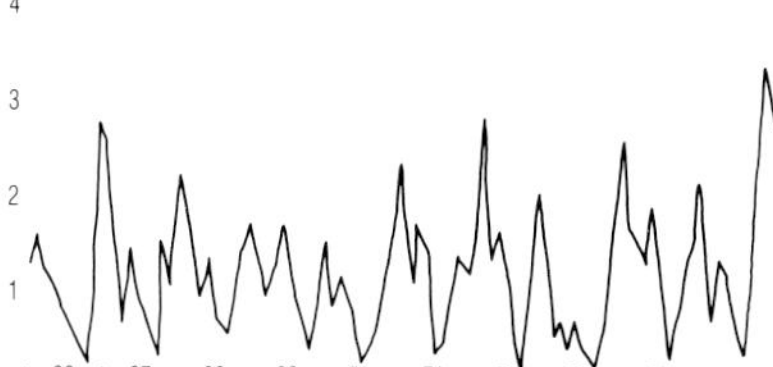

Maximum Tidal Range Meters (at Seattle)

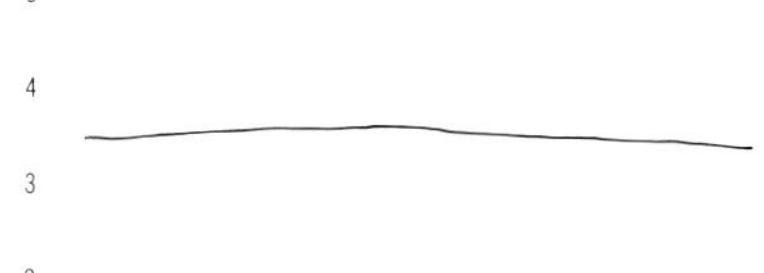

Stability Density Difference Grams/Liter Surface to 25 Meters Depth

(No Data)

Integrated Biomass Milligrams Chlorophyll *a*/Square Meter (By Month)

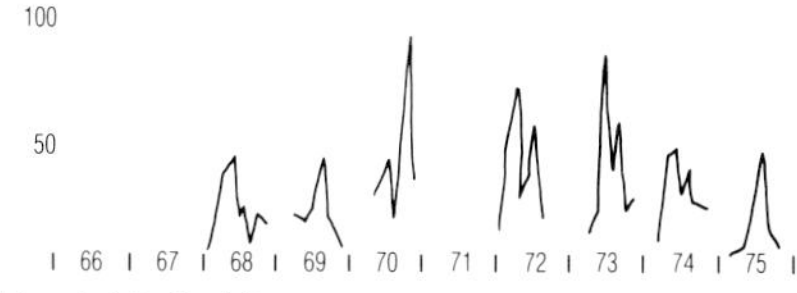

Integrated Productivity Grams Carbon/Square Meter/Day

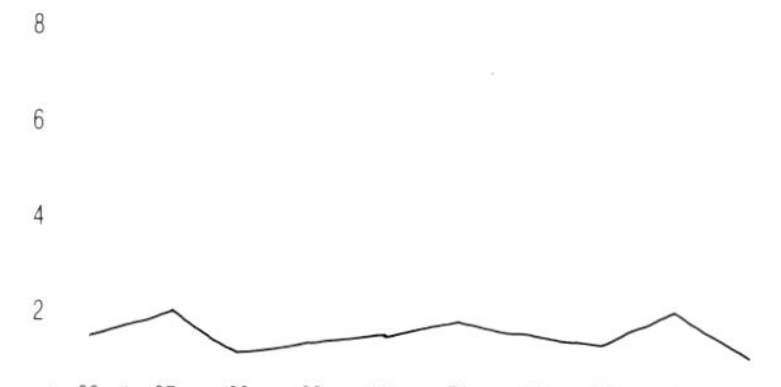

there is more sunlight after the solstice, lower production may result from increased zooplankton populations.

Figure 5.7 also shows the total annual production in the main basin for the individual years of the decade averaged above. As on land, there are good years and bad years for plant growth. The magnitude of spring blooms plays an important role in determining how productive the year will be as a whole, and there is also some relationship between a year's production and its environmental conditions.

That clearer cause-and-effect relationships are not evident in these figures is not surprising, considering the sparse quality of data. Samples were taken at a single location, at intervals of a few hours to a week, depending more on convenience than on oceanographic conditions. They are mere arbitrary grabs from amidst the chaos of oscillating stripes. While it may be overstating the case to infer that apparent blooms may just be artifacts of passing patches, nevertheless it is impracticable, by the methods used here, to completely separate spatial and temporal variability.

A case study of patchiness was made during the vigorous growth period of May 1969. A bloom, concentrated near the surface, appears around May 12 in data from single-station sampling. Surveys of surface patchiness taken through the same period, howevever, illustrate that single daily samples could be misleading, making the simple passage of a stripe resemble a bloom (Figure 5.4). There is apparently no consistent geographic pattern to these stripes as they pulse seaward in the open main basin. That is, while at a given time one spot may be greener than another, averaged over a few days or weeks all locations appear to be about equally productive. Over a period of time, the variability between locations is no larger than that at a single station.

An additional dimension of patchiness is the species composition of the phytoplankton, a fine grain beneath the gross outlines of standing stock. Surprisingly, very little information is available on phytoplankton species in the main basin. The best data (Figure 5.8), from the spring bloom of May 1967, demonstrate some important points. The dominant organisms are centric diatoms of the genera *Skeletonema*, *Thalassiosira*, and *Chaetoceros*; and populations of these organisms are extremely patchy in the main basin, fluctuating up to a thousandfold within a few days, even as total biomass varies merely by a factor of two to four.

Observations of phytoplankton species at other places and times on Puget Sound are quite scarce. In particular, there are few documented occurrences of flagellates. There is considerable evidence from other sources, however, that different phytoplankton species have very different preferences both for habitat and for environmental conditions such as temperature, light intensity, stability, and water chemistry. Di-

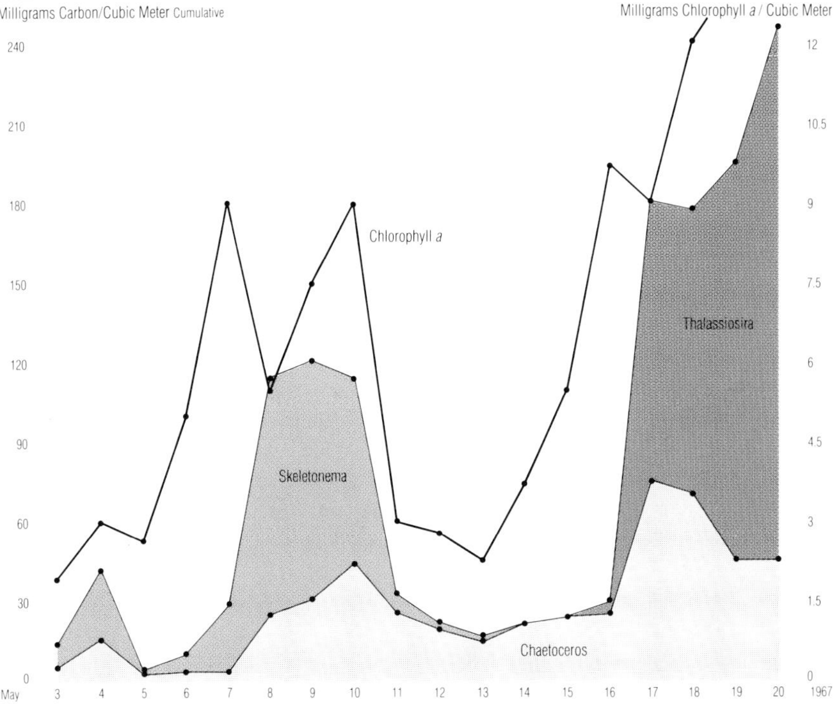

Figure 5.8 Total biomass of phytoplankton at the surface of the main basin during May 1967 is displayed as chlorophyll *a* and as carbon (estimated from cell counts of major genera). The same two blooms are evident in both sets of data, but populations vary widely: *Chaetoceros* maintained a relatively low and constant population, while *Skeletonema* bloomed after a neap tide May 5 and disappeared by the time of the *Thalassiosira* bloom at the May 20 neap tide. (After Booth, 1969)

noflagellates and phytoflagellates seem to inhabit different waters than diatoms do, being more abundant in warm, strongly stratified or poorly mixed waters. Phytoflagellates are also found in strongly mixed dark winter waters and despite their size can be the most abundant phytoplankters in both population and biomass. Observations from British Columbia suggest that while phytoflagellates may bloom at any time of year, they tend to dominate in early spring before diatom blooms have taken hold. Distinct populations of different phytoplankton groups can also coexist at different depths. Certainly the swimming ability of the flagellates must enable them to persist under conditions in which diatoms cannot remain afloat.

The importance of knowing the type of phytoplankton present, as well as the amount, will become clear when studying the food chain. Animals, including zooplankters, are selective eaters. Thus, the phytoplankton community is believed to strongly influence both the amounts and the types of zooplankton present; this in turn affects the fish community. Research on these trophic connections is just beginning to demonstrate their variability and importance.

Inlets: Variations on a Theme

The same processes that govern phytoplankton in the main basin—sunlight, mixing and residence time—operate in the rest of the Sound as well. There are variations, however, in the physical manifestations of these processes. The topography of different basins influences current speed and direction, as well as mixing and transport by winds and tides. The amount of river runoff alters stratification, residence time, and even water clarity. It is a game of ecological poker, in which each arm of the Sound is dealt a different hand from the same physiographic deck, with its biological behavior determined accordingly.

With a cautious reminder of the variability that characterizes all areas of the Sound, its subdivisions can be distinguished from each other on the basis of their average conditions of mixing and residence time. If these two variables are made the axes of a coordinate system, then basins of the Sound can be placed into quadrants that denote their physical and biological characters (Figure 5.9). There are insufficient data to make this representation strictly quantitative; rather, each subregion of the Sound is placed qualitatively along the spectra from weak to strong mixing, and from short to long residence time.

The two axes express the two measures of phytoplankton fertility. At a given intensity of sunlight (assumed, for simplicity, to affect phytoplankton uniformly over the Sound), mixing governs productivity, and residence time governs standing stock. Intermediate values of both are most favorable for sustained primary production, and thus an area such as the main basin, which on the average lies close to the intersection of the axes, has the greatest productive potential. The farther a basin is placed from the center of the coordinates, whether by increases or decreases in mixing or residence, the lower its potential production.

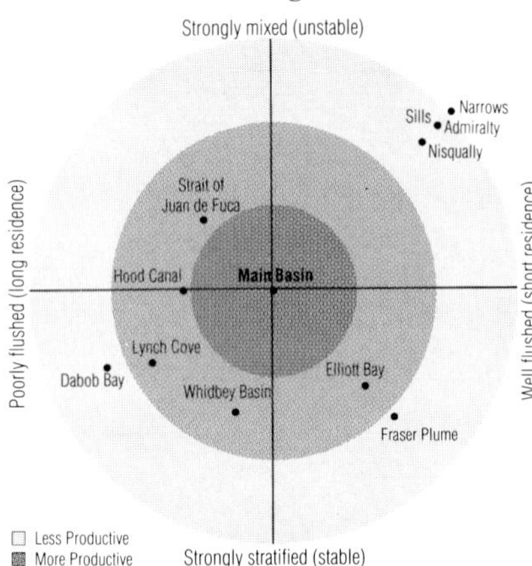

Figure 5.9 Relative productivities of Puget Sound inlets. Compared to the main basin, which seems to have a nearly optimal balance of mixing and flushing for phytoplankton growth, other areas of the Sound are less productive. They may be flushed and mixed too much (as at the sills) or not enough (as in inlets), or another combination of suboptimal conditions.

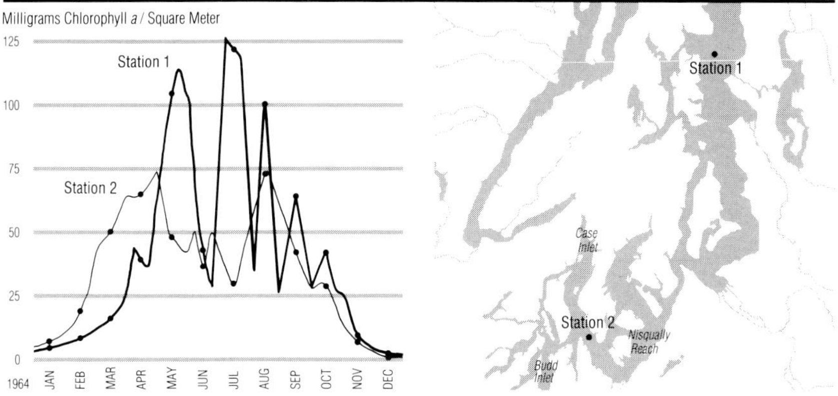

Figure 5.10 In 1964 phytoplankton biomass near a sill in the southern Sound (station 2) showed an earlier spring increase than in the main basin (station 1). During summer, standing stock was lower and less variable at station 2 due to strong vertical mixing in shallow water. (After Anderson, unpublished data)

Sills and Passages

Sills, such as those beneath The Narrows and Admiralty Inlet, are locations of strong mixing and short residence time, past which water is rapidly transported by tidal currents. There is little seasonal variation in the stability or the nutrient content of the waters over sills. Some of the best data on Puget Sound sill areas (Figure 5.10) come from a passage at the mouth of Case Inlet in the southern Sound. Compared to the main basin, stability and phytoplankton biomass change little at this location, on either the long time scale of the seasons or the shorter scale of blooms. Production here begins up to two months earlier than in the main basin, apparently because shallow water limits the depth to which mixing can occur even in the absence of stratification. This mixing restrains photosynthesis later in the season, however, and annual production at this location is estimated to be 270 grams of carbon per square meter, less than 60 percent of that in the main basin.

Similar conditions can be inferred at other sill locations. Data from the area surrounding the San Juan Islands, for example, demonstrate vertical homogeneity in the waters that seesaw through the shallows between the Straits of Georgia and Juan de Fuca.

River Mouths and Fronts

River mouths are areas of short residence time and of strong stratification and weak mixing. This is particularly true where a large river enters a sheltered inlet in which mixing forces are reduced.

There are additional reasons why river mouths can be less productive. Many rivers entering the Sound—particularly the large pristine rivers, which carry mostly snow melt rather than lowland drainage—are quite low in nutrients, further diminishing the productive capacity

Milligrams Chlorophyll *a* / Square Meter

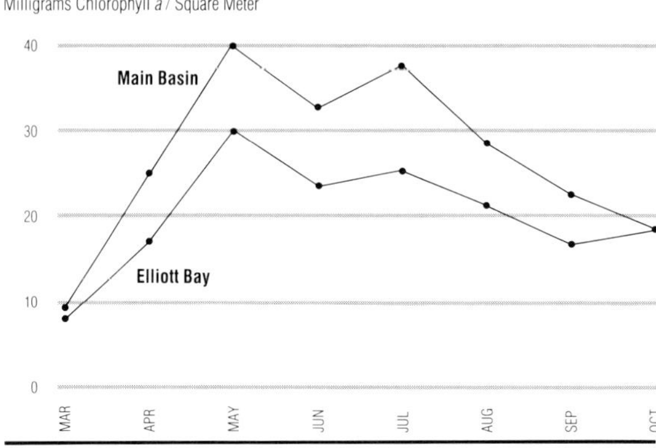

Figure 5.11 Average monthly phytoplankton biomass is consistently lower in Elliott Bay than in the main basin, despite high nutrient concentrations and strong stratification, due to turbidity and rapid flushing. (After Ebbesmeyer and Helseth, 1977)

of a stable runoff layer. Furthermore, these rivers can carry various sorts of turbid suspended matter (from natural glacial flour to urban waste), which absorb sunlight otherwise available to phytoplankton.

Elliott Bay, for example, is only about two-thirds as productive as the open main basin throughout the year (Figure 5.11). Despite its enrichment by nutrients in sewage effluent from a treatment plant in Renton, productivity at the mouth of the Duwamish River is inhibited by turbidity and by the low residence time of the outflow. During most of the year, the few kilometers of river upstream of the mouth have a low biomass of mostly freshwater phytoplankton. At the times of lowest runoff (and therefore longest residence time), usually in August, there can be intense blooms of marine diatoms as far as 10 kilometers upstream. Similar conclusions might be applied to Commencement Bay and the Puyallup estuary, which have received little study.

The same principles, plus some others, operate on a grander scale in the Whidbey basin. Saratoga Passage, between Whidbey and Camano Islands, receives the prodigious runoff from the Skagit River, which is glacial, turbid, and nutrient-poor. Substantial additional runoff from the Snohomish River enters Port Gardner near the basin mouth. The basin waters are strongly stratified, rapidly flushed, and poorly productive. The stratification is reinforced by the shelter of the island and by the lack of a shallow sill at the basin mouth. Tidal currents, furthermore, are weakened by a hairpin turn through narrow Possession Sound at the basin entrance, and by subsequent dissipation in its broader upper reaches.

Certain unique features combine, however, to make Possession Sound itself a very productive spot. Here, the spring outflow from the Whidbey basin tends to collide with that from the main basin. Reinforced by tidal intrusions and by southerly winds, this collision between waters produces a definite boundary zone or stripe called a hydrographic front, something like a weather front in the atmosphere. Phytoplankton standing stock along this front is consistently higher than that of either the main or Whidbey basins (Figure 5.12), due in part

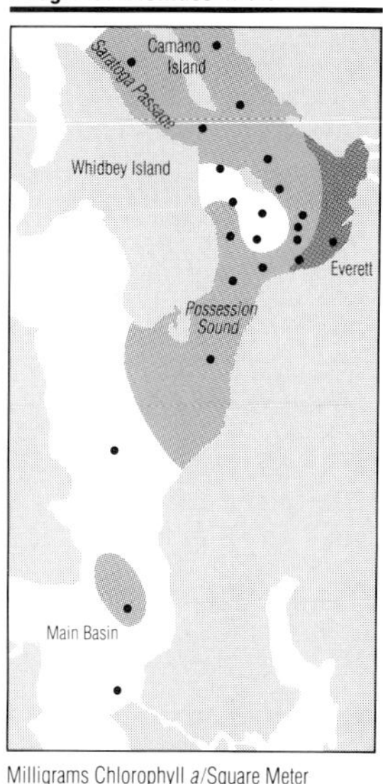

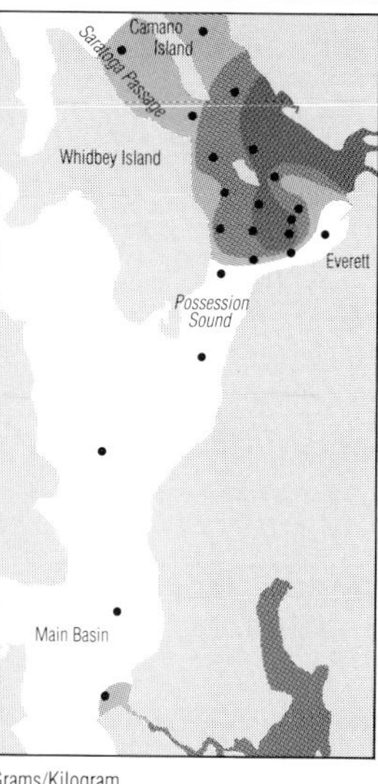

Figure 5.12 Phytoplankton abundance along a river-mouth front. Chlorophyll *a* concentrations are consistently higher in Possession Sound than in the Whidbey basin or the main basin. The highest chlorophyll concentrations in this snapshot (data from June 4–5, 1974) are found along the edge of the Snohomish River plume, where the surface salinity gradient is extremely sharp (river salinity is near zero; normal surface salinity in the main basin is 25 to 29 parts per thousand) (After English, 1979, and unpublished data).

Suspended Matter

Integrated Primary Production

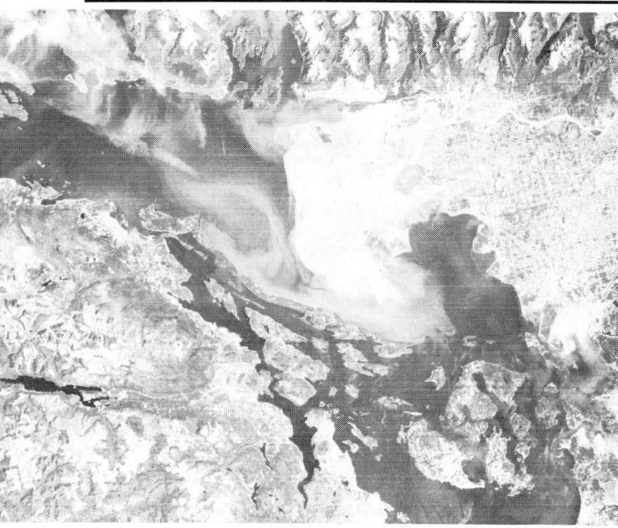

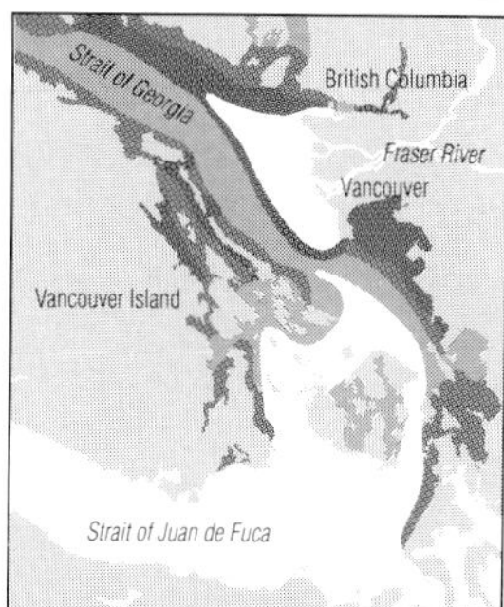

Grams Carbon/Square Meter/Year
< 200   200-300   300-400   > 400

Figure 5.13 Mean annual primary productivity in the Fraser River plume (right) is highest along the freshwater-saltwater boundary. Within the plume, salinity and nutrients are low and turbidity is high, while beyond it stratification is weak. Fraser River runoff is three to five times greater than all Puget Sound rivers combined, peaking in June due to snowmelt. High productivity also occurs in shallow, protected, nearshore waters. (After Stockner et al., 1979)

 LANDSAT photo (left) taken during ebb tide, July 20, 1974, of the 600–700 nanometer wavelength band revealing suspended matter and chlorophyll. Plumes of suspended sediment are visible off the Fraser, Nooksack, and Skagit rivers. Also visible are tidal fronts or "stripes."

to the longer residence time where the two opposing flows are stalemated. It also results from the overlapping of stable and nutrient-rich waters, each contributing half of what is needed for a bloom. Such a front may also foster the blooms in the Duwamish River. Such fronts are probably better-defined and more persistent than the smaller-scale stripes in the main basin.

An even larger and more pronounced front has been studied in the Strait of Georgia, surrounding the mouth of the Fraser River. The runoff from the Fraser exceeds that from all of Puget Sound, and spreads far across the open strait. The surface front where it contacts the surrounding seawater forms a large green ring (Figure 5.13), which oscillates with the tides and the wind.

Until enough measurements were made to reveal its existence, the green ring confused early research on waters of both the Strait of Georgia and the San Juan Islands. Phytoplankton standing stock changed radically at some locations within the few hours of a tidal change. Under a sustained northerly wind the ring could be blown amidst the islands, producing an apparent sudden bloom.

The interior of the green ring is believed to be turbid, nutrient-poor and unproductive, although some researchers suggest that it recently has been fertilized by sewage from the city of Vancouver (see Chapter Seven). That this conclusion has been disputed by other researchers highlights again the problem of drawing conclusions from limited sampling of a highly patchy environment: with the movements of the ring, only an extensive survey can place single measurements in their proper context.

Although recognized for years, such fronts have begun to receive more careful attention with the advent of rapid survey techniques. Research off England and elsewhere indicates that fronts may be highly favorable sites for the growth of dinoflagellates. Such knowledge may prove important in understanding species compositon and patchiness in Puget Sound, and may even provide clues to the outbreaks of red tides.

The Strait of Juan de Fuca also exhibits a boundary between different water masses. Estuarine water from the Sound and the Strait of Georgia flows seaward at its surface and raw deep Pacific water flows landward along its bottom, both oscillating with the tides. In addition, surface water from the Pacific can intrude into the Strait of Juan de Fuca as far east as Port Angeles, forming a front where it contacts inland surface water (Figure 5.13). These intrusions seem to be caused by the wind—not just by westerly winds in the Strait itself, but by larger-scale wind patterns over the ocean, causing currents which drive ocean water eastward into the Strait. Intrusions have been documented both by measurements of surface currents, and by the capture of species of

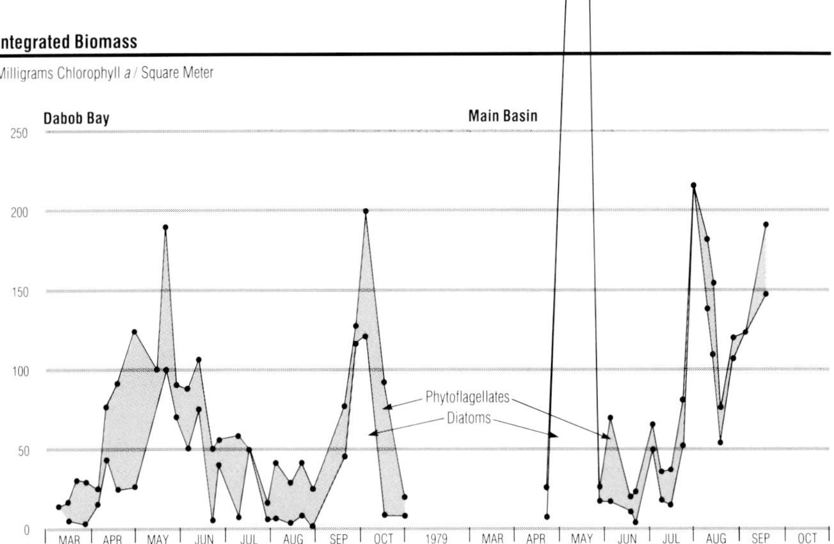

Figure 5.14 Biomass of phytoplankton smaller than five micrometers (mostly phytoflagellates) and larger phytoplankton (mostly diatoms) in Dabob Bay and the main basin in 1979. Dabob Bay began to bloom at least six weeks before the main basin, but was less productive after May, and did not bloom again until nearly October when stratification lessened. The standing stock of phytoflagellates is consistently higher in Dabob Bay, reflecting weaker mixing. (After Runge, 1981)

plankton found almost exclusively in Pacific waters. Thus the biological character of the Strait is that of a buffer zone: at its western end it is mostly oceanic, and at its eastern end it is mostly estuarine.

Deep Inlets

Three large, deep inlets in Puget Sound—Case and Carr Inlets in the southern Sound, and Dabob Bay off Hood Canal—all face south. Like the Whidbey basin, therefore, they are cul-de-sacs for surface outflow; they have weak estuarine and tidal currents, and they retain water on a southerly wind and are flushed by a northerly wind. Unlike the Whidbey basin, however, they receive little river runoff.

Such inlets thus have long residence times, and despite weak stratification are poorly mixed. While all have shallow areas at their mouths, the water motions and resultant mixing are much weaker than those of the main basin. Water is blocked from moving through the inlet—as parcels do in the main basin—by the dead-end nature of the circulation. The resulting biological pattern is one of high potential both for blooms and for subsequent surface nutrient depletion.

Such inlets tend to be productive early in the season because of their weak mixing. Later in the summer, however, and for the year as a whole, they tend to be quite unproductive due to nutrient exhaustion (Figure 5.14). In such instances, phytoplankton may be unable to grow in the upper ten meters, and the community is likely to be dominated

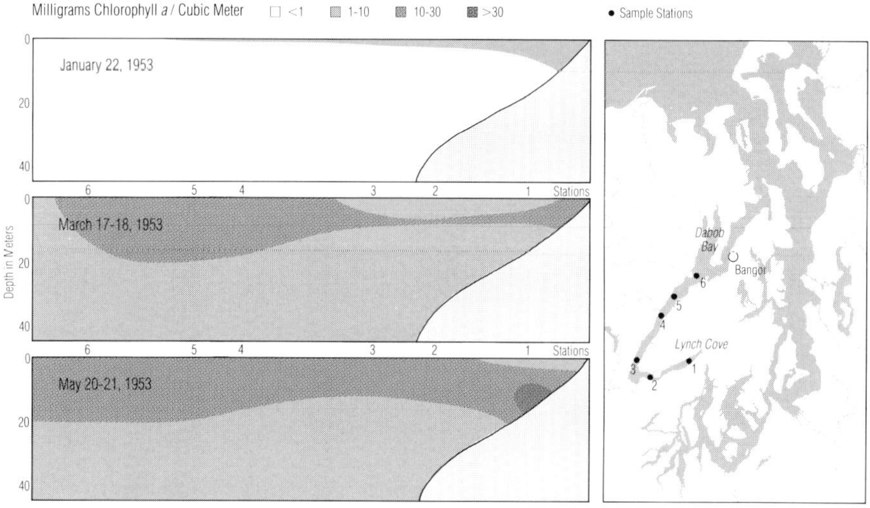

Figure 5.15 The 1953 spring bloom in Hood Canal began near the surface in the shallow waters of Lynch Cove, where vertical mixing is restricted. The bloom spread deeper and farther from shore as runoff and solar warming stratified the surface waters. As the season progressed, persistent stratification in the shallow waters caused nutrient depletion near the surface. (After Barlow, 1958)

by flagellates. The annual primary production of Dabob Bay, for instance, is only 340 grams of carbon per square meter. Unlike sill areas of comparable annual production, growth in these areas is highly intermittent; blooms are even more pronounced than those in the main basin. The weak mixing in these inlets permits a bloom to erupt during a sunny spell at any time of year, or during summer when a northerly wind flushes the stagnant surface water and leaves nutrient-rich deeper water in its place. Thus averages are even less representative of instants of time in these inlets than they are in the main basin.

Shallow Inlets

Scattered about the perimeter of the Sound are several other smaller shallow inlets: Henderson, Budd, Eld, Totten and Hammersley in the southern Sound, Lynch Cove at the head of Hood Canal, Quartermaster Harbor on Vashon Island, Sinclair and Dyes Inlets and Liberty Bay off Port Orchard, Sequim and Discovery Bays off the Strait of Juan de Fuca, and East Sound in the San Juan Islands. Most of these waters have received little study. They appear to be poorly flushed because of the baffling effect of all the passages through which water must travel to reach them, and therefore have a long residence time. Although the tidal ranges at the heads of these inlets can be larger than those elsewhere in the Sound, nevertheless the volume of water and the current speeds that accompany them are generally lower.

As observed over the sill in the southern Sound, shallow spots where the extent of vertical mixing is limited will begin to bloom early in the spring. With increased sunlight and stabilization, increased productivity spreads progressively farther from shore, as observed in Hood Canal (Figure 5.15). The open stretch of the Canal somewhat resembles the main basin in its size and orientation, but it has been studied little and is probably less productive annually because it receives half the amount of runoff of the main basin and is not enriched by an upstream sill.

Evidence suggests that many such inlets in addition to having a long residence time are also poorly mixed like their deeper neighbors. Similar results have been observed in inlets off the Strait of Georgia, including Vancouver Harbor. The most reliable indicator of late-season stratification and nutrient depletion in these areas is the appearance of harmful dinoflagellates (see Chapter Eight). Sometimes dense enough to form red tides, these dinoflagellates seem to be associated only with highly stratified waters and are scarce over sills. Additional studies have found low nutrient concentrations at the surface and low oxygen concentrations near the bottom, both signs of weak mixing and flushing.

East Sound, a shallow bay enclosed by the two arms of Orcas Island, may be an exception to the rule of low annual productivity in inlets. Blooms have been observed here in March, whereas beyond its mouth in San Juan Channel productivity was suppressed and standing stocks ten times lower until May. Productivity here may be sustained through the summer by the tidal intrusion of well-mixed water from the Channel.

Substantial gaps exist in our knowledge of phytoplankton on the Sound. Many areas have not been studied, there is little continuity in the time sequence of studies, and scales of patchiness are still poorly understood. The largest gap, however, is in the knowledge of how phytoplankton and zooplankton interact. Clearly each must have a large and vital impact on the other, yet in all the data on phytoplankton abundance there is almost no evidence of the presence of zooplankton. Apparently such effects are camouflaged by physical forces, by patchiness, and by the limitations of field research. The state of our knowledge improves only slightly when examining the influence of plants on the food chain.

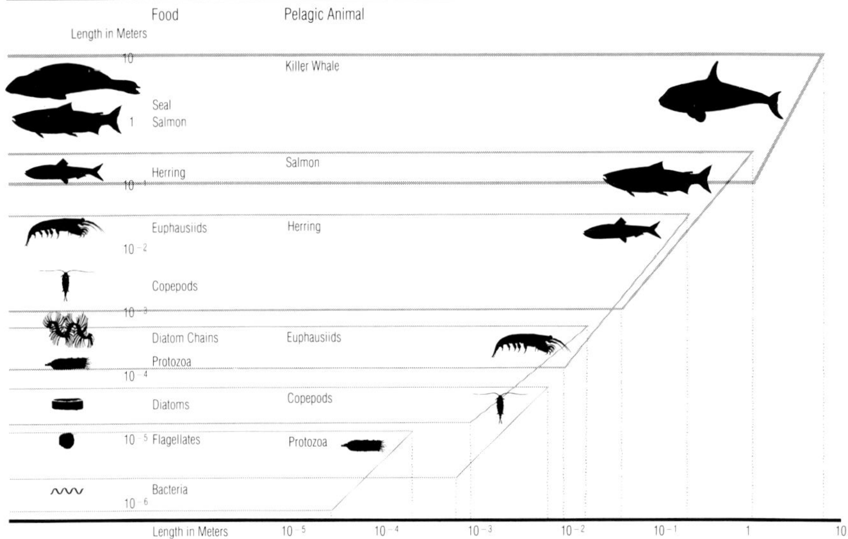

Figure 6.1 Size ranges of pelagic animals from immaturity to adulthood, and of their prey. Sizes are expressed in powers of ten on a logarithmic length scale. Length is a good measure of size in this example because predator-prey relationships often depend on length-related physical properties, such as jaw size or swimming speed. (After Dexter et al., 1981)

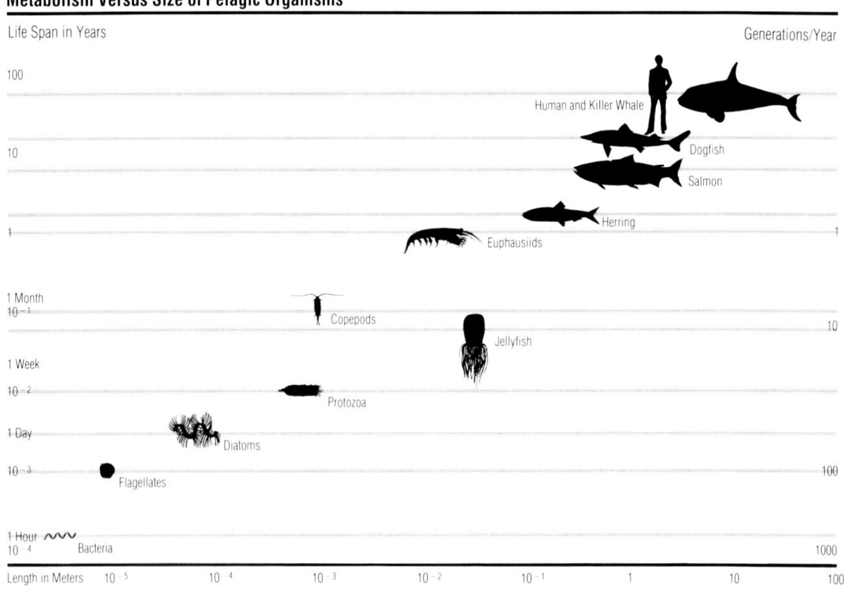

Figure 6.2 Lengths of adult pelagic organisms compared to their approximate life spans and reproductive rates, which are physiologically linked to other metabolic factors such as food intake. Metabolism is actually specified less by length than by biomass of an organism. Jellyfish are atypical in having large size but low biomass. (After Dexter et al., 1981)

66

CHAPTER SIX

The Fish Factory—Zooplankton

> ... All things are one thing and that one thing is all things—plankton, a
> shimmering phosphorescence on the sea, and the spinning planets and an
> expanding universe—all bound together by the elastic string of time.
> John Steinbeck, *The Log from the Sea of Cortez*

The term "food chain" has a mechanical ring to it, evoking images of links and sprockets, of steel and grease. Zooplankters are a vital link in this biological machinery, transforming raw vegetable matter into fish food. Yet the zooplankton community is also something of a missing link in our knowledge of Puget Sound, for less is known about it than about either the phytoplankton or the fishes. Someday it should be possible to trace the flow of trophic energy through the links of Puget Sound's pelagic food chain, but for now a general outline must suffice.

The starting point is the importance of zooplankton size in determining predator-prey relationships. Put simply, animals are larger than the organisms they eat. Predictably, the larger an animal, the larger its food. Each zooplankter actually feeds only within a characteristic range of food sizes, because it cannot capture items which are either too large or too small for its feeding apparatus to handle. The same is true of nekton. This relationship is depicted schematically in Figure 6.1.

Many idiosyncrasies are overlooked in this simplified picture. An animal may select only certain prey from the wide selection available in the appropriate size range, may discriminate between meat and vegetable food, or may occasionally violate the rule of size altogether. Furthermore, predators as disparate in size as juvenile salmon and adult baleen whales may compete for the same prey, in this case euphausiids. Nevertheless, the trend of size dependence is consistent enough to furnish valuable insights into Puget Sound's food chain.

The size of an organism—be it animal, plant, or bacterium—is also directly related to metabolism, its generation of energy from stored carbon. The metabolic rate is the fraction of body mass burned and replaced per unit of time. Again discounting slight species differences, a larger organism generally has a slower metabolism than a small one. An adult salmon, for example, eats, respires, and egests a mass of food equal to its own weight only over a span of days or weeks, while a small copepod may consume as much, proportionately, in as little as a day.

The metabolic rate in turn is related to the life span or generation time of a species. Turnover of biomass is paralleled by turnover of individuals; a slower rate of food utilization is accompanied by slower maturation and aging. Larger organisms consume less food to support each kilogram of biomass, reproduce later in life, and live longer than small organisms. This trend is portrayed schematically in Figure 6.2, which strongly resembles the previously illustrated dependence of food size on animal size.

Based on these relationships of size to diet and to life schedule, a simple mechanical analog of Puget Sound's pelagic food chain emerges. It resembles a set of sprockets for speed reduction and energy transmission, connected by chains, such as those found on a bicycle. The sprockets represent organisms, the chains represent trophic linkages, and the rotations of the machinery represent the life cycles of the organisms.

The smallest sprocket, representing the phytoplankton, is driven by solar energy through the photochemical conversion process of photosynthesis. The sun's energy arrives in daily increments, and the generation times of phytoplankters, ranging from a few hours to a few days, are tuned to this rhythm. These time periods represent the approximate rate of rotation of the first sprocket. Each successive sprocket is larger, as the body sizes of the animals it represents are larger, and the rates of rotation are correspondingly slower. Zooplankton generation times range from days to months; fish life cycles from months to a few years; and those of birds and mammals upwards to many years.

The analogy of sprockets and chains also reflects how the strategies animals must adopt to deal with prey, predators, and the vagaries of the environment are related to their life cycles. Smaller zooplankters have tremendous reproductive powers, but also have tremendous appetites. Larger zooplankters, which demand proportionately less food and are thus better able to withstand unfavorable conditions, must also take care to preserve their numbers, since their powers of repopulation are more limited. All animals must adapt to the seasonal feasts and famines of Puget Sound waters; but the strategy of a creature such as a euphausiid, which can live to see all the seasons at least once, must be very different from that of a protozoan, which may pass through several generations over the course of a single spring bloom.

A spectrum of life strategies—from steady persistence to rapid fluctuation of population—characterizes organisms in all types of ecosystems. Organisms that maintain relatively constant populations in stable environments and that have consistent, predictable life cycles are called "K-selected" (K is the symbol for the population carrying capacity of an ecosystem). In contrast, "r-selected" organisms have evolved the potential for a high population growth rate (abbreviated as r), with less attention to constancy of population numbers. The latter organisms are characterized as opportunists that can multiply rapidly to fill an ecological void; they may appear on short notice when a system becomes perturbed or unbalanced. Smaller organisms tend to be more r-selected, and larger more K-selected, although a typical species combines both strategies to its own best advantage.

The relationships of size and diet to life cycle that combine to produce our mechanical analogy also reflect two principal properties of any machine. The predators and prey of the zooplankton assemble in a characteristic way, which at any instant of time forms the structure of the food chain. The ways in which these parts interact, however, change with time as populations of various species wax and wane. Like any machine with moving parts, pelagic ecosystems must therefore be timed and tuned if they are to achieve maximum performance—there must be synchronization as well as structure, a dynamic as well as a static aspect. This synchrony is represented by the meshing life cycles of predator and prey.

Below, we will examine in more detail the zooplankters that dominate the pelagic food chain of Puget Sound, and the ways in which their life cycles interact to transmit trophic energy, not just from organism to organism, but also through space and time. Just as there is a frictional loss of energy along the drive train of a machine, there is a loss of energy along the links of the food chain. In contrast to the increasing sizes of higher animals (and of the "sprockets" used to represent them), the amount of trophic energy transmitted decreases at each successive link of the food chain, so a large output of prey is required to sustain a small production of predators. We will thus conclude by examining Puget Sound with respect to a final fundamental property of machines, the efficiency with which matter and energy are transformed and transmitted, and the ways in which efficiency is affected by changes in structure and synchronization.

Food Chain Structure

The zooplankton community is a multitude of rare and common species, temporary and permanent inhabitants, and life cycle stages from egg to corpse. Its full complexity is beyond our ability to contemplate, much less to describe, and our studies of it have the quality of random peeks through a keyhole into a crowded stadium. Animals select from a broad smorgasbord of possible foods. For simplicity we must focus on the rough structural backbone, which is built around the most common animals in Puget Sound's pelagic zone and is believed to be the principal pathway of trophic energy flow up the food chain.

Feeding on the diatoms, which dominate spring blooms, are the dominant zooplankters—the suspension-feeding crustaceans. Copepods form the largest segment of zooplankton standing stock in the main basin of Puget Sound at all times of the year. Numerically dominant are small copepods such as *Acartia*, which eats phytoflagellates and small diatoms. Dominating the biomass is the larger copepod genus

Calanus, in the main basin thought to be mostly of the species *C. pacificus*. *Calanus* mainly eats phytoplankton of intermediate size, although it sometimes captures a stray larva or protozoan.

Secondary to the copepods in importance are micronekton organisms, particularly the euphausiids, amphipods, and mysids. Euphausiids are also suspension feeders, in Puget Sound consuming mostly the largest chains of diatoms and some microzooplankton. Mysids are believed to be omnivorous, while amphipods are carnivores feeding on microzooplankton.

Although many carnivorous zooplankters reside at the third trophic level—including predatory copepods, micronekton, and gelatinous zooplankters such as chaetognaths, ctenophores, and medusae—the carnivore community appears to be dominated by the nekton. Many Puget Sound fishes—from nearshore sculpins and rockfishes to ocean-going basking sharks—derive at least part of their nutrition from zooplankton; so do some birds and the smaller baleen whales that occasionally visit. The data available indicate that the principal predators on zooplankton are pelagic fishes from roughly 50 to 200 millimeters in length. This size class includes juvenile and adult herring, smelt, sticklebacks, and sand lances, and mostly juveniles of such larger animals as salmon, cod, hake, pollock, lingcod, sablefish (black cod), and dogfish. The size class also includes juvenile and adult shrimps.

On a finer level of detail, animals select progressively larger prey as they age and increase in size. Copepod nauplii, for example, begin feeding on the smallest phytoplankton, graduating later to larger diatoms. Furthermore, most planktivorous fishes feed as larvae and juveniles, and to some extent as adults, on small epibenthic organisms that live on the sediment surface near shore. This group includes harpacticoid copepods, mysids, gammarid amphipods, shrimps, cumaceans, and polychaete worms, which are of the same size classes as the zooplankton. These prey, in turn, have consumed benthic algae, bacteria, and some dead matter amongst the silt and mud. Thus, an early root of the food chain, and an especially important one near shore and in the spring, is detritus-based rather than phytoplankton-based.

As these fishes mature they take an increasing proportion of plankton. Small juvenile fish such as herring, smelt, and pink and chum salmon eat principally copepods and crustacean larvae. Large juveniles, including coho and chinook salmon, eat both micronekton and the larvae of the smaller fishes. Swarms of micronekton, visible at the surface and called "red feed," are cues to fishermen seeking to net adult chum, pink, and sockeye salmon, which are mostly planktivorous, in contrast to the adult piscivorous (fish-eating) coho and chinook salmon, which pursue baitfishes such as herring and smelt and are more commonly caught on hook-and-line.

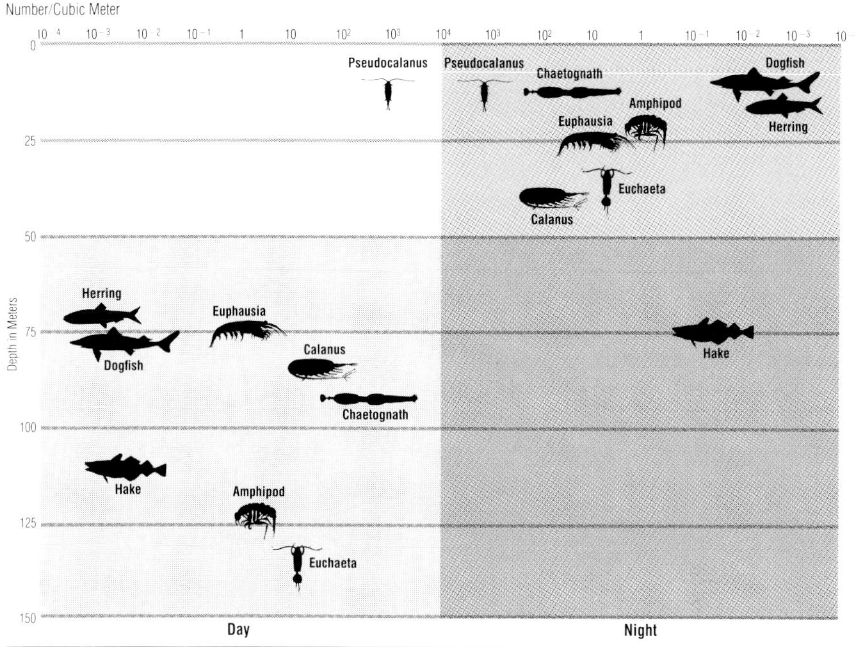

Figure 6.3 Representative standing stocks of pelagic animals (on a logarithmic scale) at their preferred day and night depths. Most animals (the small copepod *Pseudocalanus* is an exception) move to deeper water during the day. Carnivores generally are deeper and fewer, although distributions and abundances vary with time, location, and species.

This complex web can be distilled down to a simple food chain based primarily on size. The dominant diatoms of the phytoplankton feed the dominant crustacean zooplankton—copepods, euphausiids, and larvae—and these in turn feed the young and, to some extent, the adults of commercially important nekton such as herring, smelt, cod, and salmon, as well as many other non-commercial species. Of all Puget Sound's pelagic organisms, these appear to stand out and comprise what may be termed the primary food chain.

Spatial Structure

An essential detail of food chain structure in Puget Sound is the arrangement of organisms in space. Plants and animals are not distributed homogeneously in the Sound in either the vertical or horizontal dimensions, and for an animal to survive, its habitat must overlap at least partially with that of its food. Zooplankton distributions differ significantly from those of the phytoplankton, however, and seem to be determined by their physical dimensions and those of the environment.

Dominating the vertical distribution of zooplankton in the Sound is the phenomenon of vertical migration, on both daily and seasonal cycles (Figure 6.3). Migratory habits of zooplankters are correlated with

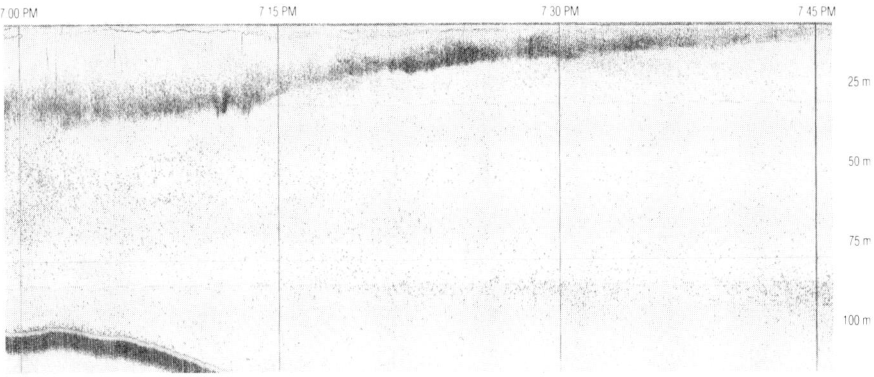

Figure 6.4 High-frequency scattering layer detected in Possession Sound near sunset in the spring. Surfaceward migration occurs over a few minutes; return to depth occurs rapidly near sunrise. (Courtesy T. S. English)

their sizes and swimming abilities. Although phytoplankton is most abundant at or near the water surface, among zooplankton the same is true only of the weakest swimmers, the microzooplankton and larvae. The rest of the zooplankton community spends much of its time below the surface—deeper during the day than at night, and deeper in the winter than in the summer. (Seasonal cycles are considered below under "Food Chain Synchronization.") The herbivorous animals feed only during their intervals near the surface.

This pattern is typical of most species of micronekton, and of roughly half of the species of mesozooplankton. Larger species, and larger (older) individuals of a given species are more likely to be migratory. Vertical migration patterns are difficult to detect in microzooplankton, small copepods (such as *Acartia*), and the juvenile stages of larger copepods (such as *Calanus*) that migrate as adults. Larger and older migrators make excursions that are greater in amplitude and deeper in the water. Larvae that first orient toward the surface foray progressively deeper and for longer periods of time as they mature. The result is a segregation of animals, according to age and size, with depth.

Migratory zooplankters congregate in narrow, discrete depth strata called sonic scattering layers that are detectable with high-frequency sonar. The layers are observed near the surface at night, and at depths approaching 200 meters during the day (Figure 6.4). Seasonal migrations can also be observed. Net tows within these scattering layers capture euphausiids, amphipods, and large copepods, animals of just the size detectable by 105 kHz sonar. Whales feeding on krill (euphausiids) use sonar of a similar frequency to locate their prey. Although sonar can sense targets as small as individual fishes, equipment to sense the small-scale distributions of zooplankters within a scattering layer is still experimental. Observed from a submersible in Saanich Inlet, how-

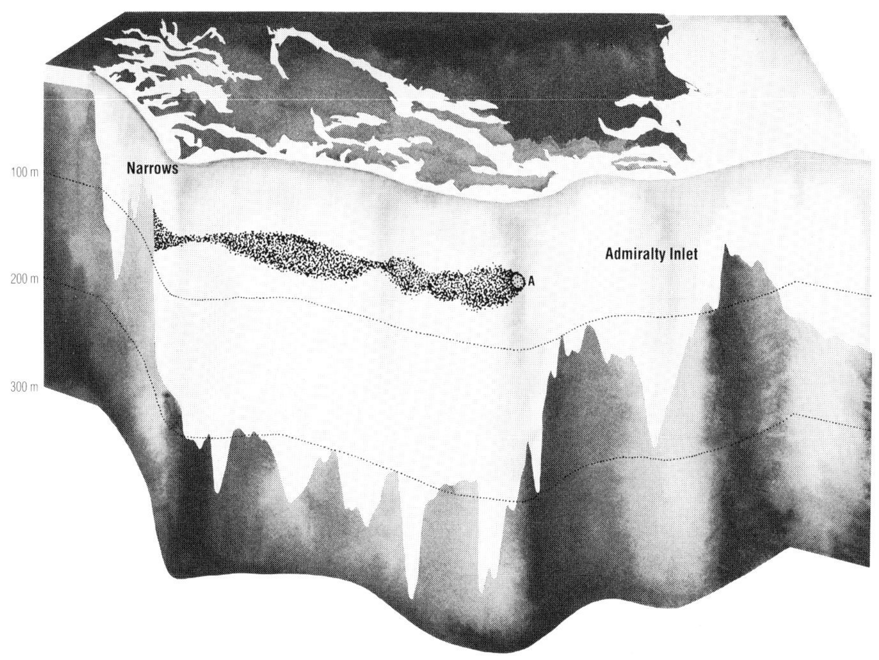

Figure 6.5 A high-frequency scattering layer is found throughout the deep waters of Puget Sound during the summer, and is absent only from shallow turbulent areas such as The Narrows and Admiralty Inlet. The layer disappears from open waters during winter (probably also due to turbulence), remaining only in protected deep inlets. "A" indicates where the scattering layer passes out of the main basin into the Whidbey basin. (After Cooney, 1971)

ever, copepods appear to be aggregated in very dense strata just a few meters thick.

Sonic scattering also yields information on horizontal zooplankton distributions. Scattering and the associated zooplankters are characteristic of deeper, quieter waters. Scattering layers are not observed at shallow locations of high turbulence, such as The Narrows (Figure 6.5). Large, migratory zooplankters may avoid these places either because of the turbulence, or because the water is too shallow to permit migration. Near shore smaller nonmigratory mesozooplankton and microzooplankton assume a more important role, due to the exclusion of the larger migrators. An extreme example is the large copepod *Neocalanus*, which spends the winter at depths exceeding 300 meters in the Strait of Georgia, cannot complete its life cycle in shallower Puget Sound, and so is found here only occasionally when it accidentally washes in.

There are more subtle differences, too, in the zooplankton fauna of various areas. More oceanic species such as the large copepods *Metridia* and *Calanus marshallae* are found in the deeper and more stratified waters of Dabob Bay and the Strait of Juan de Fuca. Certain truly offshore species are carried into the western Strait of Juan de Fuca during oceanic intrusions, the most visible of which is the floating oceanic jellyfish *Velella*, which bears a triangular sail to carry it before the wind, and which appeared in unusual quantities along the Washington coast during the spring and summer of 1981 and in the Strait of Juan de Fuca in April 1983.

The spatial arrangement of planktivorous fishes and their predators in Puget Sound closely parallels that of the zooplankton. Many fishes form tight layers detectable by sonic scattering. There is also a correspondence between fish size, depth, and distance from shore. Fish usually spawn close to—or even landward of—the edges of the Sound. Fish larvae and juveniles congregate near shore and close to the water surface. As animals age, grow, and seek larger prey, they are also found farther offshore and deeper. Herring, dogfish, sablefish, and salmon undergo daily and seasonal vertical migrations, which tend to be proportional in amplitude to age and size, and which appear to be synchronized with those of their planktonic prey. Farthest from shore and migrating most extensively are the largest animals, including such relatively rare planktivores as the gray whale.

Emerging from these data is a spatial gradient in the sizes of Puget Sound pelagic animals, in which larger species and older individuals are found farther from the shore and the surface. Thus the links and sprockets of the food chain extend from the land and the surface of Puget Sound toward the bottom and the Pacific, because energy travels bottomward and seaward from its source, the sunlight striking the Sound. It is transported by three mechanisms. The first is the passive transport of plankton and detritus by sinking to the bottom, and by currents out to sea. The second is a "bucket brigade" effect, whereby each organism is eaten by a larger one whose habitat extends deeper and farther from shore. The bucket brigade is reinforced by the third mechanism, active migrations of animals into deeper and more offshore waters as they mature, carrying inshore trophic energy with them as they go. While it thus appears that Puget Sound is a net source of energy for more offshore waters, there has been little research on the subject. Furthermore, the amount of energy transported and the times at which it is transported cannot be determined without considering the element of synchronization.

Food Chain Synchronization

The transmission of trophic energy from lower to higher organisms depends on the coupling between trophic levels. To be coupled, an animal and its food must be in the same place at the same time. The complexities of life cycles, seasonal migrations, and population variability, however, can separate predator and prey in space and time. Because

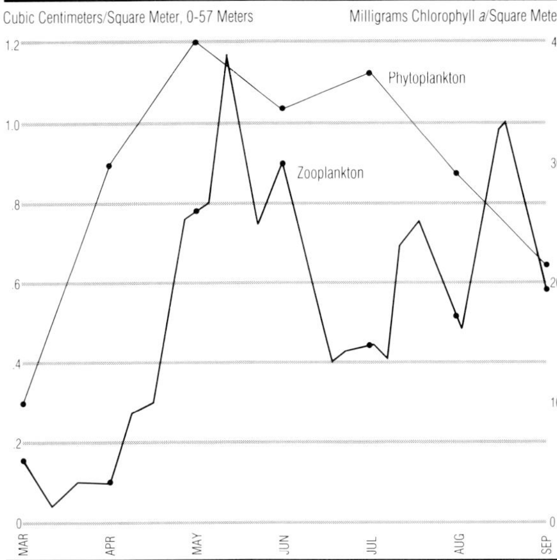

Figure 6.6 Monthly average phytoplankton abundance (Figure 5.7) compared to 1955 weekly zooplankton volume in the upper 57 meters of the main basin during daytime. The seasonal pattern must be interpreted with caution, because protozoa, micronekton, and most vertical migrators probably escaped the net (mesh size 239 micrometers), and volume could vary fourfold in 24 hours. The most abundant animals were *Pseudocalanus*, *Corycaeus*, *Calanus*, and amphipods. (After Hebard, 1956)

animal and plant sprockets are so dynamic, the food chain may frequently become uncoupled. When it does, no trophic energy is transmitted; a link is missing from the biological machinery. Food chain synchronization therefore involves the orchestration of life cycles and migrations to arrange for the maximum extent of coupling.

Life Cycles

When coupled, biomass and trophic energy are transferred from the eaten to the eater. The biomass of the organism being eaten declines (or increases at a slower rate), and that of the animal doing the eating increases. The result should resemble that depicted in Figure 6.6, which presents some of the only seasonal zooplankton biomass data available from the main basin. Peaks in zooplankton abundance appear to just follow peaks in phytoplankton, as would be expected from a coupled transfer of biomass.

This particular pretty picture, however, is likely to be almost completely fortuitous. The composite of mixed species of plants and animals hides both methodological problems and significant differences in feeding and reproductive behaviors of different zooplankters. Nevertheless, the data of Figure 6.6 suggest the adaptation of zooplankton life cycles to the long-term probabilities of food supply. These adaptations involve the timing and the frequency of reproduction.

The simplistic notion of correlated peaks in food supply and animal abundance is most easily observed in the zooplankters not represented in Figure 6.6, those with the shortest and simplest life cycles, the protozoans. An increase in food supply, which will stimulate a rise in biomass in any animal, will also soon boost the population numbers of a short-lived animal such as a protozoan. Likewise, a decline in food supply will trigger a rapid decline in population. Microzooplankton population changes can be more easily observed (as they have been in such places as Saanich Inlet and the Strait of Juan de Fuca) because they are not confounded by stages of immaturity, migratory patterns, or changes in food preference. These quick population adjustments indicate tight coupling between two trophic levels.

Uncoupling is best illustrated using salmon, at the opposite end of the pelagic size spectrum from the protozoans. When a juvenile salmon eats its zooplankton prey, its increased biomass will not be converted into new offspring until spawning, two to five years later (if the animal lives that long). During that time the fish may migrate as far as the Aleutian Islands and back, and consume other prey from benthic worms to adult herring. On this journey, the link between salmon and Puget Sound zooplankton is uncoupled, and another is forged in the open Pacific. This new link may also, at any stage, become uncoupled if, say, the salmon migrates to an unproductive area barren of prey. The difference between the salmon and the protozoan, however, is that being relatively large and mobile, the salmon can store food, and can migrate when food is scarce. The fish adjusts its behavior, more than its numbers. Populations of such animals do not fluctuate as rapidly as those of the microzooplankton; the turning of the food chain gears is smoother and steadier at this upper end of the machine.

In keeping with their sizes, microzooplankters are more r-selected, and salmon are more K-selected. The life cycles of larger animals are longer, more complex, and more rigid than those of the microzooplankton. Although salmon can maintain steady populations through temporary food shortages, the inflexibility of their life cycle prevents them from repopulating rapidly after a catastrophe, or from producing young whenever food conditions are favorable, as protozoans can. To be successful over evolutionary time, animals of all sizes must have adapted their life cycles to conform to those of their prey in some optimal fashion. That is, they must have learned to play the long-term odds on when and where to find proper food for each of their stages of maturity. At the same time, the animals must adapt to minimize their own losses to competition and predation, and their prey must do the same.

The copepods and euphausiids, with sizes and life spans intermediate between those of the protozoans and the salmon, dominate the Puget Sound zooplankton. They are adapted to the seasonal rhythms of the dominant phytoplankters. Although the smallest copepods may reproduce continuously all year, there is a general maximization of zooplankton reproductive effort during the spring bloom season. Most zooplankters reach sexual maturity and release eggs when food is most abundant, and continue feeding and reproducing throughout the summer as long as the food supply lasts. During winter when food is scarce, most zooplankters restrict their activities, conserving energy while they live off stored food, and many enter a physiological resting state (diapause) analogous to hibernation.

Within these generalizations, there is some variability. The large copepod *Neocalanus*, for example, releases its eggs deep in the Strait of Georgia in February of each year regardless of the state of phytoplankton growth. Its nauplii float to the surface, and the young of the year return to the depths by July. Puget Sound's medium-sized *C. pacificus*, in contrast, must swim to the surface and feed on the first spring blooms in order to mature and produce eggs, and continues to reproduce nearer the surface throughout the summer as food permits. *Neocalanus* produces just one generation per year; *C. pacificus*, several. *Euphausia*, while producing just one generation per year (possibly two), does not occur simply as one life stage at a time like *Neocalanus*, but rather as a mixture of stages. During the growing season euphausiids are more or less constantly maturing, producing young, and dying, but there is a dramatic increase in egg production in April and May with the first spring blooms.

Thus the mesozooplankton and the micronekton have adopted strategies of both the r-selected microzooplankton and the K-selected fishes. Their life cycles are programmed to concentrate reproductive effort when it is most likely to succeed, but they have also retained the flexibility to reproduce through most of the growing season. Their life schedules also include a rest period to maintain populations during the winter. A diversity of strategies has evolved to permit different species to coexist.

The meroplankton larvae exhibit a similar synchronization. Planktivorous larval and juvenile fishes inundate the surface of Puget Sound from roughly March, when larval herring begin feeding, through August, when most young salmon have left the Strait of Juan de Fuca. The reproductive timetables of benthic animals are not as well studied, and appear to be adapted to several factors in addition to the abundance of pelagic food, competitors, and predators. Some animals seek an optimal temperature: the Pacific Oyster, *Crassostrea gigas*, waits to spawn until the surface water reaches 21°C (70°F). Others seem to seek dispersal of their larvae: the polychaete worm *Nereis vexillosa* swims to the surface and releases its gametes in a brilliantly luminescent mating ritual at high tide on summer nights near the full moon, when tidal currents will be strongest. The limited and scattered data indicate, however, that the appearances of all planktonic larvae are strongly influenced by food availability, and allow for some adaptability to variable conditions.

Migrations

An additional dimension is added to the problem of synchronization by the bottomward and seaward movement of animals as they grow. Predators must program into their life cycles the location of each stage of maturity, as well as its timing, to couple itself to its prospective food. Some migrations are also dictated by the hydrographic conditions of Puget Sound.

During the fall and winter, for example, zooplankton populations decline both in surface waters and in the deep waters of certain poorly sheltered areas. The prominent sonic scattering layer disappears from the main basin and, slightly later, from Elliott Bay. There may be a population decline due to predation and hunger, but this disappearance seems more strongly linked to increased winter turbulence and flushing, like that found year round at The Narrows. Scattering layers persist through the winter in more sheltered locations such as Carr Inlet and the Whidbey basin. This raises the possibility that larger zooplankters undergo horizontal seasonal migrations, overwintering in the depths of protected inlets to seek shelter from turbulence and seaward currents, and thereby to conserve energy and population. This needn't tax their limited swimming powers; the landward currents in subsurface waters help carry them to shelter if they don't surface and reverse their progress. In the spring, likewise, larvae floating to the surface get a free ride back out into the main basin aboard seaward currents.

Animals overwintering in inlets are also close to waters that warm and bloom earlier in the season, making a more favorable environment for larval growth. (There may be less of an advantage in the Whidbey basin, with its heavy springtime runoff and turbidity, than in the southern Sound.) Assuming that production in inlets begins to decline by June, when production in the main basin is near its seasonal peak, riding surface currents seaward at that time would not only increase the foraging area, but would also deliver larvae to more productive waters and thus increase their rates of growth and survival. Such predictable behavior might also make them more vulnerable to predators, however.

The seasonal migrations of planktivorous fishes and their predators are better documented and parallel the apparent movements of the zooplankton. From the variability in life span and migratory locations within and between species of pelagic fishes, the trend emerges of seaward migration from the spring through the summer. Herring, for instance, begin their lives in January and February as eggs attached to eelgrass and other objects along the shores of protected inlets in the southern Sound and Port Orchard, around Vashon, Marrowstone, and the San Juan Islands, and along the southern Strait of Georgia. They spend their first summer close to the shore, then move farther offshore

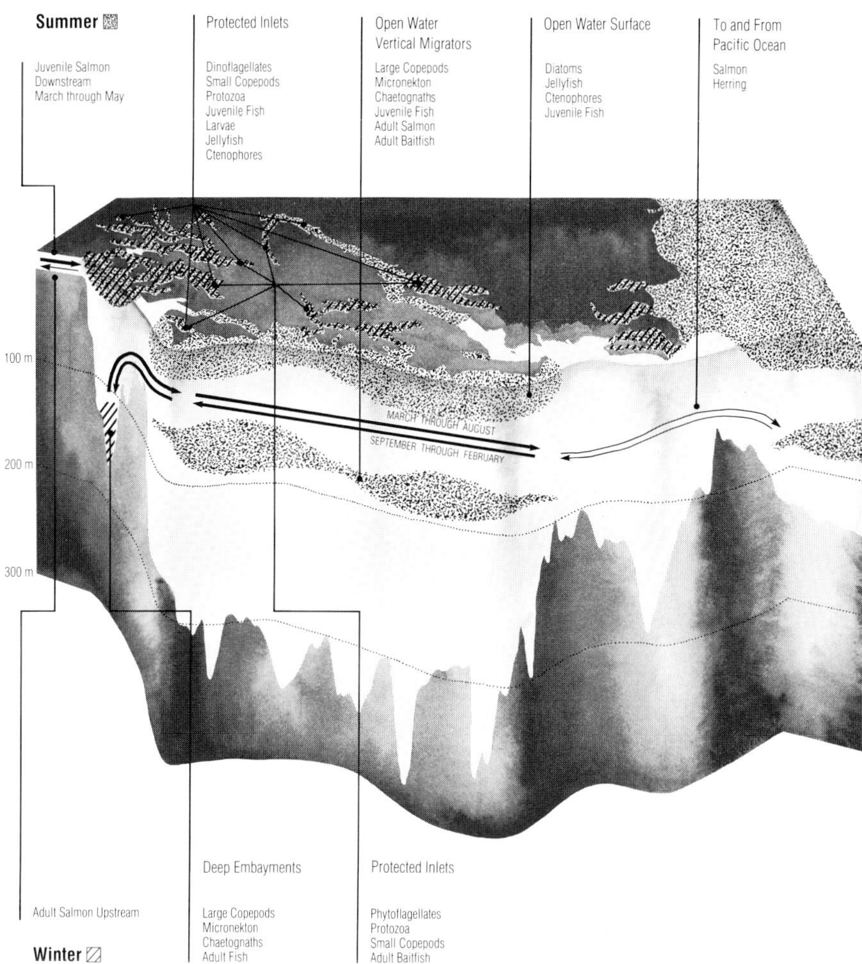

Figure 6.7 Seasonal life cycles and migrations of dominant pelagic organisms of inlets and the main basin. This caricature depicts idealized seasonal contrasts and overlooks minor organisms and variability between years. Protected inlets are sites of earlier phytoplankton blooms and are refuges for animals in the winter and spring. As deeper open waters stabilize and bloom in the late spring and summer, animals follow. The migrations reverse in the fall, completing the cycle. (After Dexter et al., 1981)

and increase in size as the season progresses. After migrating as far as the open Pacific, they return to spawn each winter and early spring near the beaches where they were hatched.

Smaller planktivorous fishes, such as the sand lance, remain closer to shore. The larger members of the cod family, while not known to migrate to the Pacific, are seen to congregate in the Whidbey basin and Case and Carr Inlets during the winter. Members of the salmon family emerge from freshwater streams as juveniles between April and June, and as they increase in size feed progressively farther from shore and deeper in the water, exiting the Strait of Juan de Fuca by late July of the summer in which they enter salt water. Resident populations of coho and chinook salmon spend their entire lives within the Sound. They are suspected of originating mostly from streams farthest from the Pacific and from the populations of juveniles which are last to reach salt water.

The life cycles and migrations of zooplankton and nekton in Puget Sound can be combined with the knowledge of phytoplankton to produce a proposed grand scheme of pelagic synchronization, depicted in Figure 6.7. It shows a coordination between the seasonal cycles of production, reproduction, and migration at all levels of the food chain. Phytoplankton production begins to increase in shallow and protected waters in the early spring, is in full swing throughout the Sound in the summer, declines in the autumn, and is virtually dormant in the winter. Phytoflagellates are present all year, but may be most important in the winter and the earliest stages of spring when other types of phytoplankton are sparse. The microzooplankton and small copepods that consume them reproduce all year near the surface and the shore. Large diatoms dominate the spring and summer blooms, and dinoflagellates perhaps make regular appearances in the late summer and fall. Diatom feeders, the large copepods, and later the euphausiids, rise from deep water to begin feeding and release their larvae in March and April. This is also the time when the zooplanktivorous larvae of herring, smelt, sand lance, codfishes, and salmon begin to arrive. Both zooplankters and their predators, like the phytoplankton blooms, increase in abundance and appear to move from inlets to the open Sound through May and June. These animals, in addition, begin to take up residence deeper in the water by day, migrating to the surface by night. By the fall, fishes that migrate to the Pacific will have already left.

The entire cycle begins to reverse itself in the fall. Phytoplankton production drops, although it can be temporarily stimulated in inlets by the renewed stirring of stratified waters, and continues longest in the shallows. Congregations of zooplankters disperse in the main basin and remain only in protected waters. Fewer larvae are around, and there is a general trend among the larger zooplankters to remain deep in the water rather than visiting the surface. Salmon return from the sea and begin heading upstream; herring and codfishes also withdraw to more sheltered waters. Herring deposit their eggs on protected beaches in January, and the cycle begins again.

This synchronized cycle is, of course, highly idealized. Many organisms do not conform exactly to it, and many other details have been left out or are still unknown. We began, however, wanting to know how trophic energy is transmitted, and we have seen that the important element is the degree of food chain coupling created by structure and synchronization in the pelagic zone. When the synchrony outlined above works according to plan the coupling is tight, and trophic energy is transferred efficiently.

The life cycle programming of pelagic animals cannot, however, always accommodate the weather-related temporal patchiness that characterizes primary production. There is a high degree of randomness to the timing and placement of phytoplankton blooms, which seldom coincide with their long-term average. What happens when animals are waiting and no bloom occurs? Or when a bloom happens prematurely, with no animals around to exploit it? Zooplankton and fishes with fortunes keyed to a semi-strict calendar can depend on the weather of a given year no more than humans can when planning their vacations.

The result of the unpredictability of nature is frequent and inevitable uncoupling of the food chain. When this occurs, links are missing from the works, and energy may be lost from the pelagic zone. Uncoupling affects another mechanical property of the pelagic food chain—its efficiency. Like the gas mileage of a car, the efficiency of an ecosystem is the yield of animals generated from the energy consumed. The tighter the coupling, the higher the efficiency; energy lost due to uncoupling reduces the efficiency.

Efficiency and the Secondary Food Chain

The efficiency of a food chain is the fraction of the original sunlight remaining when the raw materials of carbon dioxide and water are converted to their final animal product. Whatever the degree of coupling, significant energy losses occur at each stage of transformation. Calculated values of food chain efficiency, though imprecise, never exceed a mere fraction of a percent. The greatest loss occurs before the biological processes are triggered, in the transit from the sun to the phytoplankton. Both incoming solar energy, and the trophic energy it becomes when stored in biomass, can be measured in kilocalories, the same unit of energy used to measure dietary requirements. Of the roughly twenty kilocalories of sunlight entering each square meter of the atmosphere during each minute of daylight, at most one-third reaches Puget Sound. Figure 6.8 shows that less than one percent of the energy reaching the upper atmosphere is actually absorbed by phytoplankton to serve as the starting point for the pelagic food chain.

Once metabolic processes begin, there is a different set of losses: plants and animals use some of their biomass as fuel simply to keep their metabolic furnaces burning, and expend additional to energy for locomotion, reproduction, etc. Of the light energy absorbed by phytoplankton, less than two percent is stored more than temporarily in plant matter. The rest is expended to power the photosynthetic appara-

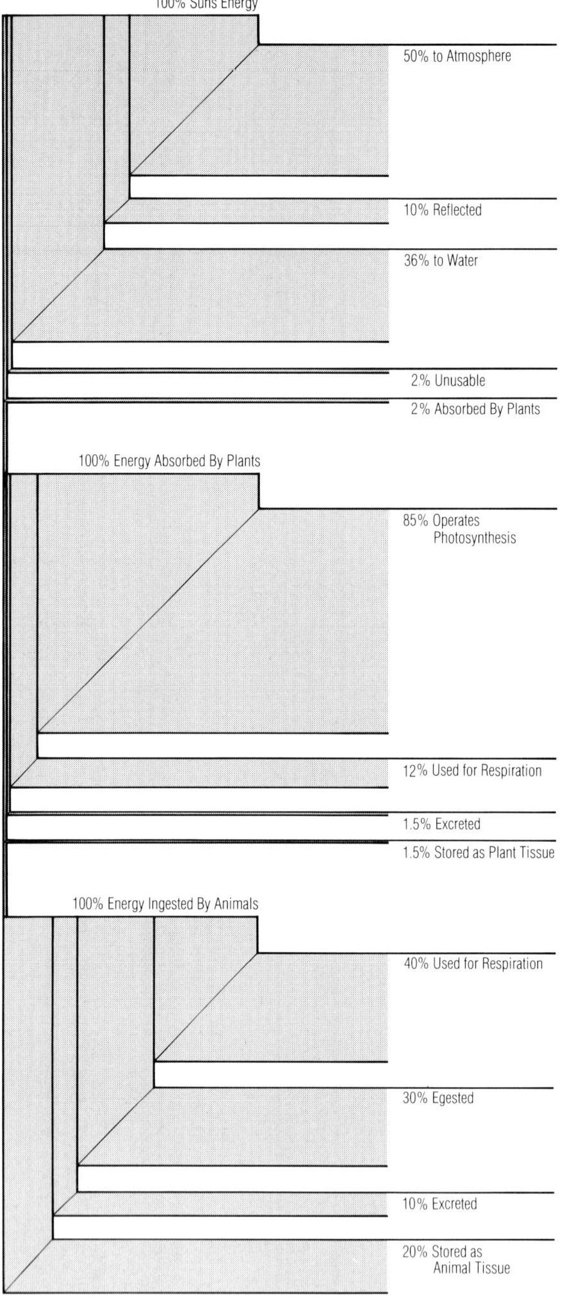

Figure 6.8 Only a tiny fraction of the solar energy reaching the earth enters the pelagic food chain. Night hours exclude one to two-thirds of sunlight (depending on season); nearly one-third is removed by scattering and absorption by air and clouds; and some is reflected off water. Of the solar energy that penetrates Puget Sound, half is of the wrong wavelength to stimulate photosynthesis, and most of the rest is absorbed by water or detritus instead of by plants.

Of the solar energy that is absorbed by phytoplankton, about 85 percent drives photosynthetic reactions and 12 percent powers other metabolic activities. Less than two percent remains as plant tissue to be eaten by animals.

Of the solar energy stored in plants that are eaten by herbivores, roughly 40 percent is voided as feces and urine, and another 40 percent is metabolized. About 20 percent is left to be eaten by carnivores. Similar losses apply at successive trophic levels.

tus, to acquire the nutrients and vitamins necessary for building tissue, to operate the hereditary works in the nucleus, to excrete, or to move. Of the matter eaten by an animal, a major portion is egested as feces rather than assimilated; most of the rest is respired or excreted, and only a fraction is added permanently as animal tissue.

The numbers in Figure 6.8 are crude averages, which vary greatly among different organisms and conditions, and are difficult to measure reliably. A rule of thumb in terrestrial ecosystems has been that when one organism is eaten by another, about 90 percent of the material is metabolized or eliminated. The productivity of an herbivore, for example, is thought to be only a tenth that of the plant on which it feeds. The "transfer" efficiency—the fraction of energy transmitted from one trophic level to the next—is therefore traditionally reckoned at 10 percent. The transmission of energy to a third trophic level, then, is only one percent, and to a fourth only a tenth of a percent.

Steele has argued that in the ocean, the transfer efficiency may approach 20 percent rather than 10 percent. In reality, its value is probably quite variable, and the lesson remains that the food chain is an inherently costly and inefficient machine. The difference between 10 and 20 percent, however, is highly significant when the multiple links in the food chain are considered. The productivity of a carnivore on the third trophic level, for example, would differ by a factor of four, on the fourth trophic level, by a factor of eight, and so on.

Furthermore, regardless of precise values of efficiency, another factor governs the overall efficiency of an ecosystem, the length of the food chain between primary producers and human harvest. The more links in the food chain between raw material and product—for example, between plant and fish—the lower will be the cumulative efficiency, and the lower the yield. Ryther has hypothesized, for example, that coastal upwelling areas, such as that off Peru, are the richest fish-producing areas of the world partly because there are fewer trophic levels. A shorter food chain, the theory says, minimizes metabolic losses and maximizes harvest. If Puget Sound salmon could feed one link lower on the food chain, by this reasoning, they could expand their potential food supply, populations, and yield to humans, by a factor of five or ten.

The Secondary Food Chain: Hypothesis

Greve and Parsons have expanded on Ryther's theory in studying the relationship between efficiency and food chain coupling. They suggest that paralleling the primary diatom-crustacean-nekton food chain in temperate coastal waters, such as those of the Puget Sound area, is a secondary food chain that supplants it under certain conditions, and which is less efficient at producing commercially important fishes.

Populating this other side of the ecological tracks are some of the smallest plankters—the phytoflagellates, protozoans and small copepods, and some of the largest—the carnivorous ctenophores and medusae. When diatom populations dwindle under unfavorable environmental conditions, such as strong stratification (or possibly pollution), the primary food chain is uncoupled, and trophic energy is theorized to be diverted instead to this secondary food chain.

Because this secondary food chain originates with smaller phytoplankton, three trophic levels rather than two are required to produce zooplankters of the sizes preferred by larval and juvenile fishes. Ryther's theory, therefore, would predict a five-to-tenfold decrease in the fish yield from this lengthening of the food chain.

Greve and Parsons, however, further argue that fishes eat few ctenophores and medusae, making the secondary food chain a dead end that shunts trophic energy away from fish production altogether. There is some persuasive evidence to support these contentions, but the complexity of the pelagic realm presents alternative explanations as well, which merit a closer and more detailed examination.

The plankters of the secondary food chain are highly r-selected, with short life spans and the potential for rapid growth under favorable conditions. Their abrupt appearances and disappearances have been characterized as a boom-or-bust pattern of abundance. The gelatinous ctenophores and medusae, or "jellies," have a much higher body-water content than do the crustaceans, and do not fall neatly into the pattern of life span versus length (Figure 6.1). They can reproduce rapidly when food is abundant, and are much shorter-lived than other zooplankters of similar dimension. This discrepancy is rectified if animals are compared on the basis of the dry masses, which among the gelatinous carnivores are similar to those of microzooplankton. Other zooplankters potentially classed on the secondary food chain are the larvaceans and the chaetognaths, which possess the dimensions of micronekton but the mass of a small copepod.

The carnivores of the secondary food chain also capture prey differently than those of the primary food chain. Ctenophores and medusae sweep their prey from the water passively using sticky cells and stinging cells, respectively. Neither of these are raptorial predators, which select their prey individually, as carnivorous crustaceans and vertebrates are thought to do. When food is abundant, gelatinous carnivores can process large volumes of water and gather more prey than a raptor; but when prey are sparse, selective hunting requires less energy and raptors have an advantage. The result of this combination of vigorous reproduction and feeding is that gelatinous zooplankters can appear suddenly, devour huge quantities of small prey, then just as suddenly disappear. They are reported to accumulate, almost like red tides, into dense windrows at certain times and places in the summer, but are virtually absent during the winter.

A major consequence of this r-selected ecological niche is a higher metabolic rate for a given-sized animal. More of the calories ingested by these animals are expended in respiration, and thus the transfer efficiency among such animals may be at the lower end of the assumed 10 to 20 percent range.

Although it is difficult to verify because their remains are more difficult to identify in fish stomachs than those of crustaceans, there is also justification for the belief that gelatinous zooplankton are an inferior food source for fishes of all ages. Being mostly water, they yield little nutrition for their size, or for the energy expended to capture them. The sticky or stinging cells may also discourage predators.

Thus there are three theoretical reasons why, if a distinct secondary food chain does exist, it should be less productive: it is longer, making for more metabolic loss; it incurs greater metabolic losses at each link; and it produces a food of poor quality for commercially important fishes. What this theory does not take into account, however, is the fate of biomass and energy when the food chain is uncoupled. It overlooks the fundamental problem of the fluid world: that food which cannot be stored in living organisms sinks to the bottom and is lost to the pelagic zone almost entirely.

The Secondary Food Chain: Field Evidence

The theory of Greve and Parsons attempts to explain the results of the CEPEX experiments in Saanich Inlet (see Chapter Seven), in which water and plankton were captured in enormous flexible plastic cylinders (9.5 meters in diameter and 23.5 meters deep) suspended from the surface. The reduction of water motion in the cylinders apparently caused diatoms to sink out, and flagellates to replace them. (Greve and Parsons proposed that this replacement can also be triggered by pollution.) In such instances, large copepods were observed to die off, and to be replaced by jellies; juvenile salmon placed in the enclosures appeared to be slowly starving, and trophic energy was apparently shunted into a dead end.

What happens to this energy when the primary food chain is uncoupled? Its fate is dictated in part by the physiological differences between the zooplankters of the primary and secondary food chains. The urine and feces of both types of animals contain high concentrations of the nutrients (especially nitrogen) that were originally mixed upward from deep water to be taken up by phytoplankton. As long as these waste products remain near the surface, they can help replenish the nutrient supply through biochemical decomposition and recycling by bacteria, and so help sustain phytoplankton growth. Much of the excreta sinks out of the surface, however, and comes to rest permanently on the bottom, where it sustains the benthic community. Only a small amount of this "lost" matter and energy, yet to be measured, returns to the pelagic zone through vertical mixing, and via surface-seeking meroplankton larvae.

Protozoans and gelatinous zooplankters release their feces in amorphous form, as marine "snow," which sinks slowly and fragments as it goes. Crustaceans release larger, compacted, membrane-enclosed fecal pellets, which sink at rates of tens of meters per day. Larger animals also live deeper in the water, so their excretion is of less value to surface-living phytoplankton, especially if it occurs under conditions of high stability and poor vertical mixing. As a result, the larger the zooplankton, the more rapidly matter is transported bottomward—because of faster sinking of feces, and because of vertical migration—and so the greater the fraction of trophic energy that escapes from the pelagic zone and reaches the benthos.

These differences have been measured in Dabob Bay. Sinking suspended matter was collected in sediment traps, cylinders several centimeters in diameter, hung vertically on a cable at several depths below the surface, and open at the top until retrieval. Most phytoplankton growth in Dabob Bay occurs in the spring and early summer. Most of the grazing was also observed at this time, as reflected by the appearance in the traps of digested chlorophyll voided in fecal pellets. But most of the carbon did not fall to the bottom until autumn, having been carried about in the bodies of zooplankton all summer. Little material reached the traps during phytoflagellate blooms; instead, it apparently was retained in the surface layer where it could refuel phytoplankton growth. Under the tight coupling between phytoflagellates and microzooplankters, not only did animal waste sink more slowly, but less of the phytoplankton sank out uneaten. During diatom blooms, however, large quantities of raw or poorly digested plant material sank bottomward. In such cases, diatom growth must have been uncoupled or poorly coupled, since only with minimal grazing could bloom populations have appeared and sunk out. A greater likelihood of uncoupling, in fact, is expected from the less flexible, more K-selected organisms of the primary food chain.

It thus appears that the blooms so characteristic of phytoplankton in Puget Sound (especially of the diatoms of the primary food chain) are in fact evidence of uncoupling—of animal populations insufficient to consume plant matter at the rate it is generated. The magnitude of spring blooms can be attributed in part to the presence of only the few adult herbivores surviving the winter; grazing pressure does not become intense until the first hordes of copepod and euphausiid larvae are produced from the final feeding of their parents. This contrasts with

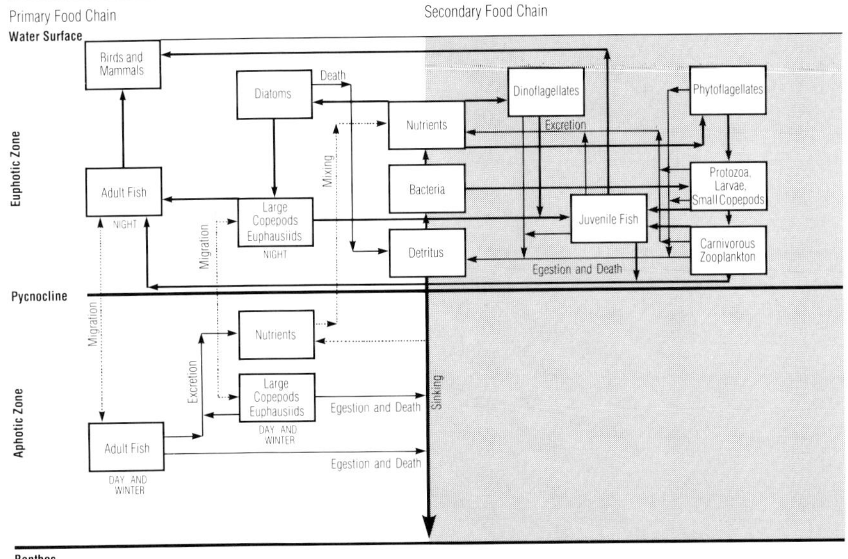

Figure 6.9 Simplified diagram of the pelagic food web of Puget Sound, incorporating the primary and secondary food chains. Many secondary food chain organisms are important foods for animals of the primary food chain. The secondary branch of the food web also is characterized by less sinking and migration into deep water, and by tighter coupling and more surface nutrient recycling, all of which are advantageous when environmental conditions are unfavorable.

conditions in the Gulf of Alaska, where *Neocalanus* nauplii, released surfaceward early in the spring by the unfed adults, begin feeding at the earliest opportunity. The tight coupling in these waters effectively maintains phytoplankton standing stocks at a constant level throughout the year, despite a spring increase in primary productivity.

The paradoxical result of these differences, is that, coupled or uncoupled, the main, diatom-based track of the food chain appears to deliver large quantities of trophic energy to the bottom, as well as to the fishes. The delivery may be direct, by sinking when uncoupled, or indirect, through the bucket brigade of zooplankton, fish, and their waste products. Larger quantities of energy may be transmitted by diatoms than by flagellates, but apparently more of it also is transported into deep water, where it is at least temporarily lost to surface organisms. In a location such as Dabob Bay, which in contrast to the main basin suffers stratification and poor vertical mixing, this can lead to a serious decline in productivity.

The secondary food chain, on the other hand, while seemingly less productive, is more tightly coupled and retains organic and recycled nutrients near the surface. In the few areas of the ocean where such research has been conducted—the Grand Banks, the Gulf of Mexico, the Bering Sea—it appears that areas dominated by flagellates and microzooplankton may actually produce more pelagic fishes, and those populated by diatoms and copepods more bottom fishes. Thus the so-called secondary food chain may be neither less efficient than the primary food chain, nor even distinct from it.

The most realistic, yet still comprehensible, picture of trophic relationships in Puget Sound eliminates the artificial distinction between primary and secondary food chains. They are actually entangled with one another, branches of a more complex food web, as depicted in Figure 6.9.

Bridging the gaps between the two food chains are the changes in size and feeding habits of animals of the primary food web branch, which both pass through and are dependent on the secondary branch. Dinoflagellates and the protozoans, larvae, and small copepods that consume phytoflagellates are themselves prime food sources for larval fishes. Fishes and other animals are very sensitive to the availability of the proper food during their first days and weeks of life, and the health of individual organisms and of fish populations as a whole depends on this synchronization. Thus it may be misguided to view the secondary branch of the Puget Sound pelagic food web as a dead end, especially in the early spring when diatom blooms have yet to begin, and when hungry larval fishes are emerging in search of a first small meal.

Flagellates in Puget Sound exhibit a conservative strategy, growing under hydrographic conditions—low light intensity, chronic stratification, depleted nutrients—that are unfavorable for the production of diatoms. Rather than displacing diatoms through competition, flagellates might be viewed as filling an ecological void left when diatoms cannot grow. Rather than supplanting an efficient food chain with a wasteful one, flagellates may supplement primary production under harsh conditions to which diatoms are poorly adapted. The organisms with the shortest life cycles and the fastest growth rates are well-adapted to respond to the changeability of Puget Sound, where favorable conditions can appear and disappear within matters of hours or days. In the trophic machinery, the secondary branch of the food web seems to take on the role of shock absorber, damping out plunges in productivity and keeping the gears running more smoothly.

CHAPTER SEVEN

Plankton and Pollution

Those who will not understand science, become instead its prisoners.
Charles A. Lowe

The final step in studying plankton is to understand how plankton and people interact. Civilization faces a great dilemma with respect to the sea: while being subjected to an uncontrolled experiment on the effects of waste disposal, the oceans are also looked to as a food source for unborn millions. We demand more of the ocean's biological wealth, while we burden it with more of our refuse. Both plankton and Puget Sound are at the center of this paradox, because both are battlegrounds for the conflicting demands of food supply and waste disposal.

Plankton plays a pivotal role in the responses of the ocean to pollution. Plankton is itself directly affected by pollution, but it can also act as either a conduit to or a shield for higher animals. The plankton is a gateway between the sea and the land, by which food is transported to people, and through which waste travels in return.

Puget Sound is a microcosm of these relationships. Here, in one place, is both a bountiful food supply with great potential for expansion, and a convenient dumping ground for the waste of human populations and industry. Today parts of Puget Sound are as pristine as forested wilderness, and parts are as developed and degraded as urban sprawl. There is no simple reconciliation of the needs for both more development and less impact. The best hope lies in combining an objective, scientific attitude with a spiritual appreciation of our relationship with the sea.

Human impacts on the sea have received intense publicity in recent years, and have aroused public indignation due to a number of catastrophes. Examples come easily to mind. Oil spills like the *Torrey Canyon* in 1967 off England and the *Amoco Cadiz* off France in 1978 blackened miles of shoreline and wounded local economies. Consumption of mercury-tainted seafood at Minamata, Japan, produced a wave of human death, illness, and birth defects through the 1950s and 1960s. Many species of marine mammals—whales, fur seals, and sea lions—in the Pacific Ocean and worldwide have been slaughtered to near extinction. The temporary decline in populations of seabirds off southern California led, as much as any other factor, to the banning of DDT for most uses in the United States. Attitudes toward exploitation of the sea have changed dramatically in the last two decades, along with attitudes toward resource use and pollution in general. Americans have reached a point where no actions can be taken, on sea or on land, without careful scrutiny of their possible effects on the environment.

But tracking down sources of pollution, determining the nature and extent of their effects, and prescribing cures can be complex tasks. A pollutant may be chemical or thermal; it may originate from as specific a source as a factory, or as diffuse a source as ocean-going vessels. Tons of one pollutant might be safer than teaspoonfuls of another. Some apparently serious threats are not so serious, and some of the worst menaces may be unpublicized and virtually invisible. The effects of a pollutant may appear far from its source, in some subtle ecological disguise. Each pollutant behaves in a unique fashion, and must be evaluated individually.

Science, therefore, has established a protocol—a common, agreed-upon set of rules by which all the pollutants may be judged, and their hazards determined. It involves gathering as much information as practicable about certain aspects of pollutants before passing judgment. The rules of the protocol are as follows:

Know the background level of the pollutant. Practically any substance, even water, is harmful to marine organisms in excessive quantities. Conversely, however, apparently any pollutant can be tolerated in a small enough dose. Certain pollutants are entirely natural or even essential substances, such as nutrients and minerals, which human activities have concentrated to an abnormal degree. These have consequences very different from those of anthropogenic substances, which owe their very existence to civilization. The hazard posed by addition of a substance to the sea depends on the amount already present, and on the degree to which human activities would alter that amount.

The concentrations of pollutants in water are commonly measured by the fraction of weight they contribute when mixed with water, in the same fashion as the salinity. Salt is present in Puget Sound seawater at concentrations of two to three percent by weight, or 20 to 30 parts salt per thousand parts water. Nutrients such as inorganic nitrogen are usually present at the surface of the Sound at 100 or more parts per billion, or 100 micrograms in a kilogram (one liter) of water. The ranges of concentrations of pollutants that are harmful to marine organisms can vary all the way from the parts per hundred (tens of grams per liter) down to the low parts per trillion (billionths of a gram, or nanograms, per liter). It does not take much of a compound to make up such low concentrations. At its present concentration of PCBs of roughly two parts per trillion, for instance, the approximately 2×10^{14} (200 trillion) liters of water in Puget Sound contain only about 400 kilograms (880 pounds) of material.

Know the sources and sinks, and their rates. A sink is the opposite of a source; it is a pathway of removal, a final resting place. No constituent of seawater is static; all are constantly being added and subtracted by various mechanisms. The relatively constant salinity of the sea, for example, is sustained by a dynamic balance of minerals constantly washing from land and settling to the sediments, and of fresh water cycling between land, air, and sea. Any accounting of human alterations must consider many natural and potential rates of supply and removal, as well as existing background levels.

Know the pathways between source and sink. Most pollutants in the sea ultimately end up on the bottom, but they might reach their destination by many routes. When determining the biological effects of such substances, the intermediate fates between source and sink must be traced, lest some important impact be overlooked, or the relative importance of various pathways be misunderstood.

Know the biological interactions. The effects of pollutants on living things are as individual and varied as the chemicals and organisms involved, and it often takes years of research to pinpoint the who, where, why, and how of their impacts. Nevertheless, some generalizations can be made. It is important, first of all, to distinguish whether a pollutant reaches an animal through direct uptake from the water, or through trophic uptake from its food. The latter is potentially far more serious, since it is the mechanism by which biomagnification, the progressive elevation of pollutant concentrations in the tissues of animals at successively higher trophic levels, takes place. Secondly, we must distinguish between acute and chronic, or lethal and sublethal effects. The immediate and obvious effects of toxins (including death) are dangerous; but because they may be delayed, or are too subtle to notice, long-term effects of small quantities of pollutants may be even more sinister. In addition, the effect of one pollutant may be modified by the presence of others. This synergism between pollutants—for example, the heightened sensitivity of an animal to one pollutant when it is already fighting the effects of another—can occur in numerous combinations. On the positive side, however, individual animals can acquire a tolerance to low levels of pollutants after a period of exposure. Furthermore, many plants and animals have the ability to depurate pollutants—that is, to eliminate them from their systems once placed in clean water—or to metabolize them and break them down into harmless by-products.

There are three classes of interactions between plankton and pollution. First, pollutants of various kinds have direct effects—usually toxic but sometimes stimulatory—on planktonic plants and animals. Second, plankters in return influence the physical and chemical conditions of pollutants in seawater, and can significantly affect their ultimate fate. Finally, of perhaps greatest interest is the possible role of plankton in the biomagnification of pollutants and in pollution effects both on the Puget Sound food web and on humans.

Effects of Pollution on Plankton

The first step in tracing the fates and effects of pollutants is to examine their direct impacts on plankton. Figure 7.1 presents the concentration ranges of some pollutants in Puget Sound, compared to minimum values found to harm plankton. Toxicity varies widely, depending on the pollutant, and the values displayed can be misleading if not properly interpreted. Most of the concentrations of dissolved pollutants in Puget Sound are at the low ends of the ranges presented, with just a few polluted sites (e.g., lead at a dredge spoil site in Elliott Bay, mercury in Bellingham Bay, arsenic off Tacoma) providing the

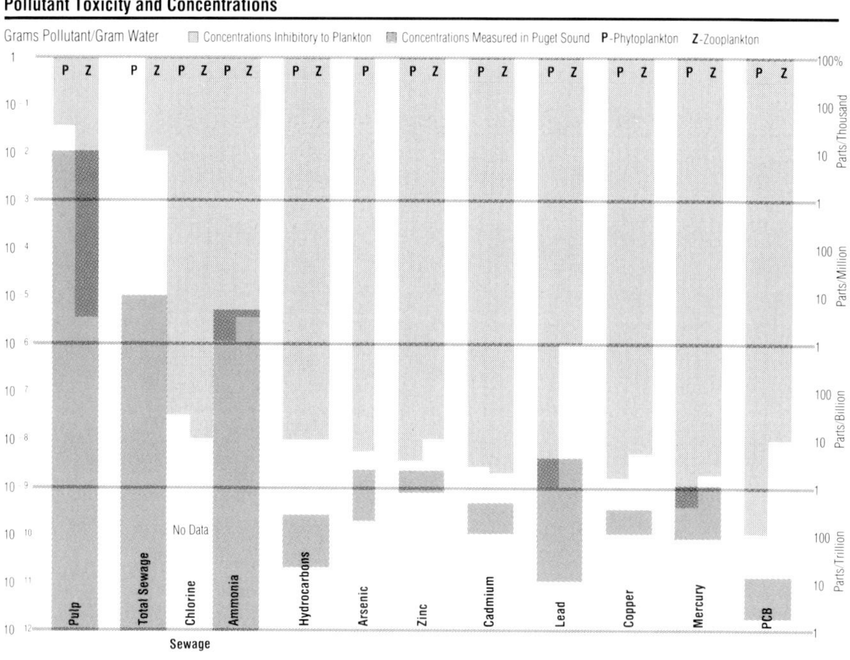

Figure 7.1 Measured concentration ranges of several pollutants dissolved in Puget Sound waters, compared to concentration ranges shown to be hazardous to plankton. Concentrations are displayed as a fraction of water mass on a logarithmic scale. Overlap of the two ranges for a given pollutant indicates potential damage to plankton. Concentrations of dissolved pollutants in most regions of Puget Sound are near background levels, with just a few polluted sites (e.g., lead at a dredge spoil site, mercury in Bellingham Bay, arsenic in Commencement Bay) providing the high values. Some of these measurements were made before recent cleanup efforts. Likewise, the concentration thresholds at which plankton is harmed are set by a few sensitive species, and most plankters are hardier.

high values. Many of the minimum unhealthy values of pollutant concentrations are for particularly sensitive species; the same values may not affect or may even stimulate other species. Furthermore, the logarithmic concentration scale shows increments of tenfold, making differences appear smaller than they would on a normal arithmetic scale. In general, the wider the range in concentration of a pollutant in the Sound, the greater the human input; this is particularly evident for such pollutants as lead.

Figure 7.1 shows that pollutant levels in Puget Sound waters are mostly below toxic thresholds. There are few documented cases of pollutant injury to plankton here, and recent research suggests that toxicity is more closely related to pollutant burden within the plankton itself than to that of the waters. The major categories of pollutants are discussed individually below, roughly in order of increasing toxicity to plankton.

Sewage

Least toxic of the pollutants—in fact, stimulatory under most conditions—is sewage. Until the 1960s, much of the waste of the Seattle metropolitan area went into Lake Washington, with results similar to those in many other bodies of fresh water that received the same abuse. The lake became eutrophic: cloudier, greener, fouled by algae, and filling in prematurely. The lake was overfertilized by the nutrients in detergents and human waste, especially phosphorus. Since the 1960s, Lake Washington has begun to return to its original state, because the sewage has instead been diverted into Puget Sound. The obvious question then is, can the same thing happen to Puget Sound that happened to Lake Washington?

Municipal sewage, before being discharged into the Sound, currently receives what is called primary treatment: the waste water is screened, held in ponds to allow much of the solid sludge to settle out, and disinfected with a spurt of chlorine. Federal regulations require sewage to undergo secondary treatment, in which the effluent is further retained and filtered to allow bacteria to digest sludge. This process can remove 95 percent of the solid matter in the effluent, although 90 percent is a more customary figure for discharge to salt water. The effluent that enters salt water from either treatment process is rich in both organic and inorganic forms of two elements, nitrogen and phosphorus, which promote the growth of phytoplankton. Tertiary treatment, used on effluent discharged to fresh water, involves removal of these nutrients as well.

Before its effects can be known, it must be determined whether sewage effluent significantly alters the chemical composition of the Sound's waters. In the case of much of Seattle's sewage, the answer appears to be no. Most bodies of salt water are more sensitive to changes in nitrogen supply than to that of phosphorus. They are also better flushed, and so effluent is better dispersed than in most lakes. Puget Sound already has an abundant background level of nutrients, which in the main basin is seldom low enough to limit phytoplankton growth. Vigorous currents and vertical mixing serve both to replenish surface nutrient supplies and to disperse and dilute sewage effluent. The natural flushing of nutrients in and out of this portion of the Sound far exceeds the amount added by people. Effluent from the city's West Point treatment plant is released near the bottom, at a depth of about 70 meters, where nutrients are abundant and the effluent is unlikely to affect plankton. Likewise, waste from the Renton treatment plant (which already receives secondary treatment) appears not to affect the frequency of phytoplankton blooms near the mouth of the Duwamish River, into which it is discharged.

In contrast, there have been noticeable increases in nutrient levels in the Fraser River plume (Strait of Georgia) during the 1970s. The sewage of the city of Vancouver is released near the mouth of the Fraser River. The enriched river runoff forms an enormous, stable surface layer which occupies the middle of the Strait. Nutrient concentrations in the plume were formerly quite low, because of poor mixing with deeper water; now, for the same reason, the nutrients persist, and they may have altered phytoplankton growth in that area.

Excessive phytoplankton growth has become a problem in several areas of Puget Sound into which sewage is released, including Sinclair Inlet and Elliott Bay. Any organic detritus—marine snow, dead plankton, or sewage sludge—will stimulate the growth of bacteria and the consumption of oxygen, and so is said to carry a Biological Oxygen Demand (BOD). While the removal of sludge from sewage reduces the BOD of the waste itself, the remaining nutrients foster phytoplankton blooms, and hence increase the potential BOD. In stratified Budd Inlet off Olympia, the remains of surface blooms sink and are trapped in denser waters near the bottom. Here they decay, consuming enough oxygen to make the water inhospitable and potentially lethal for fish and benthic organisms. In the Duwamish River, furthermore, while sewage may not affect the timing of blooms, it is suspected of supporting an increased phytoplankton standing stock, thus exacerbating a similar problem of decay. The problem arises in these locations because the deeper water is not flushed as thoroughly as in the open main basin, where mixing with surface water at sills maintains a high oxygen level. A similar oxygen shortage troubles the waters off New York City, and may be aggravated by sewage dumping.

The organic matter in sewage effluent may have additional, more subtle effects on water chemistry. Many of the compounds can chelate heavy metals, and may be altering the species composition of phytoplankton in some areas (see page 101). There is also concern about another element in sewage, the dissolved chlorine gas added as disinfectant at levels of about one part per million. Plenty of chlorine is already present in the Sound as sodium chloride (salt), and the chlorine gas dissolved in sewage effluent is diluted 140-fold before it is discharged into the water. Yet doubts persist because both chlorine and its sibling halogen, bromine, are suspected to combine with organic compounds in sewage to form organochlorines and organobromines, which are toxic to plankton at far lower concentrations than the elements alone.

Perhaps the most serious chemical insult inflicted on Puget Sound from sewage outfalls comes from the quantities of other pollutants that are dumped, accidentally and otherwise, down sewers. Metals, petroleum, and synthetic organic chemicals are present in high concentrations in municipal sewage, and are not always effectively removed by treatment. These compounds may also reach the Sound through storm sewers, which can empty urban runoff directly into both Lake Washington and Puget Sound, without even primary treatment, during heavy rains. These pollutants are discussed individually below.

In 1981, METRO, the agency responsible for King County's sewage disposal, was told by the Washington Department of Ecology that the Duwamish River could no longer handle the effluent from the secondary treatment plant in Renton. Already contributing 25 percent, it was feared that sewage effluent could increase to 50 percent of the river's volume during the late-summer low-flow season as a result of population growth east of Lake Washington. METRO planned to bypass the river, where water quality was a problem, and pipe effluent directly into the Sound at a deepwater site off Seahurst Park between Burien and Vashon Island. The impact of such disposal of sewage effluent depends on the flushing rate of the discharge area. Critics disagreed with METRO's siting choice, advocating instead a costlier but possibly better-flushed site near Duwamish Head. METRO hired consultants for an extensive study of the subject, and also obtained a waiver of the federal requirement for secondary treatment of other existing saltwater discharges, intending instead to spend its money on preventing discharge of untreated storm sewage.

Pulp and Paper Wastes

The earliest serious pollutant on Puget Sound was the effluent from pulp and paper mills, once scattered from the southern Sound near Shelton to Everett, Anacortes, Bellingham, and Port Angeles. Two processes are used to digest wood fiber chemically into pulp: the kraft process generates large quantities of sodium hydroxide (lye), sodium sulfate, and sodium sulfide, while the sulfite process releases calcium, ammonium, or magnesium bisulfite. Except for sulfide, found only in anoxic water, the quantities of these chemicals present naturally in seawater are not significantly altered by the pulp effluent.

Accompanying them, however, are dissolved organic compounds leached from the wood, including organic acids and their salts (similar in composition to soap), sugars, and lignins. These can have a number of effects. The leachate can be chemically poisonous either directly by acidifying the water, or by combining, like sewage, to form chelated or organochlorine complexes. The leachate can also deplete the oxygen content of seawater. It consumes oxygen directly by chemical action, and although some leached compounds, as well as some detergents also present in pulp effluent, may stimulate phytoplankton growth at low concentrations, the deep brown color of kraft effluent is suspected of suppressing the oxygenating effects of photosynthesis in seawater into which it is discharged. These effects are detectable only at effluent concentrations of a few percent or more, and though such effects have been observed, notably in British Columbia, pulp effluent nevertheless is one of the least toxic of pollutants.

Pulp and paper wastes are problems on Puget Sound only in locations where they have accumulated due to poor mixing. This unfortunately has been the case at most mills. A plant at Anacortes discharges into a well-flushed channel, but at Port Gardner (Everett), Bellingham Bay, and even off Port Angeles where circulation is restricted by Ediz Hook and Dungeness Spit, waste lingers close to its source. Under state government orders, the problems have been reduced since the early 1970s by chemical treatment, reduction in the volume of effluent, and transfer of release points to better-mixed locations.

The most visible effect of pulp mill effluent on plankton in Puget Sound has been the reduction in zooplankton populations in affected areas. Animals such as euphausiids and juvenile and adult fishes (especially the migratory salmon) avoid areas like Everett Harbor and Port Gardner where effluent concentrations are high. They do return, however, when conditions improve.

The worst damage inflicted by pulp waste apparently strikes animal larvae, especially oyster larvae. The once-rich oyster beds in the neighborhood of Port Gardner have shrunk since pulping began. Oysters and their larvae are at a disadvantage in accommodating pollution, as are many plankters, partly because they cannot avoid tainted areas. The toxicity of Port Gardner surface waters to oyster larvae has dropped in recent years, which may herald a recovery. But the example is a reminder that not all species are equally hardy. Larvae—especially oyster larvae—are among the organisms most susceptible to all types of pollutant stress. For this reason, scientists evaluate the toxicity of waters in Puget Sound by studying their effects on larval oysters, in a procedure called a bioassay. The marine equivalent of laboratory mice, larvae are grown side-by-side in clean water and in water to be tested for pollutant effects, of which deformity or death of the plankters is a measure.

Petroleum

Petroleum pollution of seashores is highly publicized, but its effects on plankton have received less attention. Crude oil is a complex and highly variable mixture of liquid hydrocarbons, each with a different chemical structure and a different weight, grading all the way from light, volatile gasoline to heavy tar. The components that separate during the process of refining also separate when crude oil is spilled on water. What we see on birds and beaches is the heavy fraction, most of which eventually sinks to the bottom. An unseen fraction, containing the lighter hydrocarbons (especially the aromatic hydrocarbons related to benzene and toluene) is far more toxic to marine life. Crude oils from different locations vary in composition and toxicity, but most of the petroleum products refined for people's use—gasoline, lubricating oil, diesel and home heating fuel—are rich in the light fraction, and so are more toxic than plain crude oil. When spilled onto seawater, much of this lighter fraction evaporates, sinks to the bottom, or is decomposed by bacteria, but depending on the conditions—wind, waves, etc.—some of it also dissolves. The invisible dissolved compounds pose the principal threat to plankton.

The major sources of petroleum input to the seas are also nearly invisible. The highly publicized tanker accidents (which Puget Sound fortunately has been spared so far), together with the flushing of bilges, account for no more than a third of the petroleum entering the world's oceans. Nearly the same amount is suspected of entering the sea naturally, through submarine seeps.

Puget Sound is primarily affected by invisible sources of petroleum. Over half of the petroleum that eventually reaches the sea was originally discharged on land as unburned petroleum from automobiles and furnaces, and as domestic, municipal, and industrial waste. Although figures are unavailable on petroleum inputs to Puget Sound, over two-thirds of the oil used in the state of Washington ends up spread on roads, dripped from cars, dumped onto the ground, and carried by rainfall into sewers, rivers, and lakes, and into the Sound. Sewers cannot separate oil from waste water. Half of all the hydrocarbons entering Lake Washington run directly off streets and bridges, amounting to nearly 30 metric tons (33 long tons) a year. Much of the input comes during storms, when some runoff is discharged, untreated, through storm sewers. In Lake Washington and elsewhere, the overwhelming proportion of this petroleum is apparently automobile crankcase oil. The real villains of oil pollution are thus ordinary citizens. As yet, however, no effects of such inputs on Puget Sound plankton have been documented.

Numerous laboratory studies have demonstrated the toxicity of petroleum components to phytoplankton, zooplankton, and larvae, although low levels of hydrocarbons can stimulate phytoplankton growth. The results of field studies of plankton populations near oil spills (Table 7.1) are ambiguous: changes are difficult to detect, and long-term depletion is rarely observed. Although plankters cannot swim away from a contaminated area, as fish can, in most areas their abundance and regenerative capacity are apparently sufficient for rapid recovery. Spills can be devastating, however, for larvae of animals that spawn only once a year, and especially for fish larvae that concentrate at the surface where oil slicks linger. Nevertheless, the difficulty of drawing such conclusions given the inherent quality of oceanographic data must be emphasized. Special care must be taken to distinguish oil from natural plankton hydrocarbons. An interesting sidelight (discussed below) is the possible important role of plankton in removing oil from surface waters to its final resting place on the bottom.

Heavy Elements

All the elements found in nature—and now a few that have been created by technology—are present in the ocean. The composition of natural chemicals dissolved in seawater has, so far as science can determine, reached a steady state in which processes of removal to the sediments just balance the input from rivers. Certain of these elements are harmful in small doses, especially heavy metals such as mercury, cadmium, silver, nickel, lead, arsenic, copper, chromium, and zinc (in decreasing order of toxicity). Some—copper and zinc, as well as iron, magnesium, manganese, molybdenum, and cobalt—are essential to life in small doses, but in higher doses are toxic.

Although the natural concentrations of these elements in seawater are low, the sea is so huge and the turnover of elements so vast that (with the possible exception of lead) it is difficult for humans to add significant amounts. People can make their presence felt, however, by creating new elements, or by dumping large quantities of an element into a small, poorly mixed region, in a form not easily dispersed. Although the levels of heavy elements in Puget Sound as a whole are still very close to the background levels present in the open sea, some potentially harmful metals can be found in elevated concentrations at certain locations (Figure 7.1).

Heavy metals illustrate well that sources and fates of pollutants must be known in order to gauge their effects on plankton. A principal human source of metals in Puget Sound, for example, is the American

Table 7.1 Observed effects of oil spills on plankton. Although data on the impacts of spilled oil on plankton are scarce, they suggest greater resistance and powers of regeneration in the plankton than among plants and animals of the seashore where oil is trapped.

Spill	Phytoplankton	Zooplankton	Source
Torrey Canyon England, 1967	Some mortality	No data	Malins (1977)
	No data	Fish eggs and larvae mortality	GESAMP (1977)
Santa Barbara, 1969	No mortality observed	No data	Middleditch (1981)
Florida Buzzards Bay, Massachusetts, 1969	No data	Reduction in crab larvae survival for several years	Krebs and Burns (1977)
Refinery Seto Inland Sea, Japan, 1974	No visible effect	No visible effect	American Petroleum Inst. (1979)
Argo Merchant Cape Cod, 1976	No data	Reduction in biomass	Malins (1977)
	No data	Contamination	American Petroleum Inst.(1979)
Sansinena explosion Los Angeles, 1976	Temporary depletion, physiological stress	Temporary depletion, species changes	Geyer (1980)
Tsesis Swedish Baltic, 1977	Increased biomass, probably due to reduced grazing	Temporary local drop in biomass, heavy contamination	Kineman et al. (1980)
Ekofisk North Sea, 1977	Little effect	No data	Lannergren (1978)
	No data	Contaminated	Mackie et al. (1978)
Amoco Cadiz Brittany, 1978	No data	Mortality highest and recovery slowest near shore	American Petroleum Inst. (1979)
	No data	Chronic depletion in places	Spooner (1978)
IXTOC Gulf of Mexico, 1979-80	Large blooms	Mortality	Jernelov and Linden (1981)
Tidal test pond, Mississippi coast	Primary production drops 50%, recovery in 2 months	Immediate mortality, recovery within 6 months	Brown (1980)

Smelting and Refining Company (ASARCO) smelter in Tacoma. From this plant, arsenic, a by-product of copper smelting, is recovered and sold commercially. It is also dumped into Commencement Bay as crystalline slag and discharged as arsenite dust into the atmosphere and as liquid into the Sound. Arsenic is harmful to organisms because it can masquerade as the essential nutrient, phosphorus. The effects of the smelter on the Sound, however, are less than they might appear. The crystalline slag is poorly soluble in seawater, and the air- and water-borne arsenic, together with a lesser amount of natural arsenic which enters the Sound via river runoff, is very soluble. The strong mixing and flushing action of the Sound rapidly disperses the human input, and dissolved arsenic levels are elevated above background levels only in the immediate vicinity of the smelter and in the Tacoma tideflats, where slag was used for landfill.

A different pattern is exhibited by the metal mercury. While relatively innocuous in its elemental form as a liquid metal, mercury in ionic form (and particularly compounded in an organic molecule) is a potent toxin. The "Minamata Disease," a complex of neurological symptoms including visual and cognitive impairment, paralysis, and birth defects, was named for a Japanese bay from which local cats, birds, and humans consumed large quantities of seafood contaminated by factory discharge of mercury. Dangerous toxicity resulted when this mercury was compounded into an organic form.

Until 1970, a chlor-alkali plant on Bellingham Bay released five to ten kilograms of liquid (metallic) mercury per day as a by-product of the manufacture of chlorine gas and lye. This amount far exceeded the bay's natural inputs of dissolved ionic mercury, a few ounces per day, from rivers and the atmosphere. Flushing carried another five to ten kilograms per day of dissolved mercury in and out of the bay, the natural background in seawater. At their peak, dissolved mercury concentrations reached one part per billion near the outfall, but were diluted back to background levels of ten parts per trillion by the time water exited from the bay. Most of the mercury adhered instead to particles, and sank to the bottom, where much of it remains and conversion to organic form can occur. Thus, there was less of a threat to plankton than to benthic animals. Similar problems with mercury have occurred in the Strait of Georgia, but neither condition has approached the seriousness of the situation at Minamata.

Another source of metals in Puget Sound is sewage. Peak metal concentrations, caused by intermittent dumping of waste into the sewage system, have reached 800 parts per billion of copper, 100 parts per billion of lead, and 9 parts per billion of cadmium, in undiluted effluent from the West Point Treatment Plant, 10 to 100 times their background levels in the Sound. Cadmium and lead concentrations in effluent exceed safety standards specified by the Environmental Protection Agency. The potential threat is reduced, however, by dilution of the effluent before discharge and by vigorous flushing at the discharge site.

Extreme caution must be exercised when using such numbers to pinpoint possible environmental threats. In Puget Sound, the mass of metals in sewage effluent is so small compared to that in the natural seawater flow in the Sound, and to other inputs of metals to the Sound, that levels of dissolved metals show little variation within the Sound, or between the Sound and offshore waters. Technology for measuring metals is also evolving rapidly, and in some cases previous measurements have been a hundredfold too high. Furthermore, the strong interaction of metals with suspended particles of all kinds, including plankton, affects their presence in water and in organisms, and contributes to their ultimate deposition in the sediments. There is evidence, in fact, that particulate rather than dissolved metal concentrations control toxicity to plankton.

Some paradoxes and confusion over the possible effects of metals are linked to the chemical states of the elements in question. Different workers have found natural zinc concentrations in the North Atlantic to be alternately insufficient and excessive for phytoplankton growth, and there is also speculation that natural concentrations of copper may be inhibitory. Dissolved metals usually reach salt water in the form of compounds, and the nature of a compound will influence both its fate in seawater and its toxicity to plankton. Like mercury, lead is more toxic as an organic compound. Excess lead in Puget Sound comes mainly from automobile exhaust, in a tetraethyl compound. Transformations of metals into compounds of differing toxicity can be mediated by plankton. Effects of a single metal can also be complicated by the presence of others, a common situation in polluted environments.

Metallic ions in water can also couple themselves loosely and reversibly to complex organic molecules in an association known as chelation. In this state they are apparently less free to interact with other chemicals or with organisms, and their biological effects will depend in part on the amount remaining free in solution. Among the organic chemicals that may act as chelators are those in soil runoff, sewage, and pulp mill effluent, as well as some compounds released by plankters themselves. Methods to measure chelation so far are unreliable. There has been speculation that red tide phytoplankters are more sensitive to metal pollution, and that red tides are increasing in frequency because chelators in sewage protect the organisms from inhibitory amounts of metals present naturally in the sea. Much evidence conflicts with this hypothesis, however; the complexity of the effects of pollutants on phytoplankton species will be demonstrated when the CEPEX food-web experiments and red tides are examined (Chapters Seven and Eight).

Heavy radioactive elements from nuclear reactions (from bomb tests and power plants) have been judged one of the worst potential ocean pollutants. Although some radionuclides are relatively abundant in the sea, others (some of which have an affinity for living tissue) are rare and could be significantly elevated—at least locally—by human activities. Considerable study has been devoted to the coastal waters off Washington and Oregon where radionuclides from the Hanford nuclear reservation are delivered by the Columbia River. Some of this radioactivity may spread northward into ocean water that enters the depths of Puget Sound, but as yet no harmful impact has been observed either off the coast or here.

Synthetic Organic Chemicals

Perhaps the most insidious pollutants in Puget Sound belong to a highly diverse group of chemicals composed of rings and chains of carbon atoms. They resemble some of the compounds in petroleum, but differ in being entirely anthropogenic (manmade), and having been nonexistent scarcely a generation ago. Their production has skyrocketed in the last few decades, and traces of such chemicals can now be found everywhere in the world, even in the Antarctic ice cap. Up to three million such chemicals are now in commercial production, but most scientists are aware of the names—much less the biological hazards—of only a few.

One of these, valued at first for its toxicity, is DDT, banned for use in the United States since 1970 after it was publicly implicated in reproductive mortalities of such birds as the brown pelican and osprey. DDT became the classic example of a persistent biocide undergoing biomagnification. Locally, such chemicals may be implicated in harbor seal pup mortalities in the southern Sound.

Receiving less publicity, however, are perhaps thousands of related compounds that are released inadvertently to the environment and washed down, ultimately, to such places as Puget Sound. Many of these are poorly studied and difficult to recognize in the marine environment. Some attention has been paid to the effects in Puget Sound of one family of organic chemicals, the polychlorinated biphenyls (PCBs). Belonging to that class of substances known as halocarbons or organochlorines (in which we also find the by-products of sewage chlorination), PCBs are highly toxic and persistent and make a useful case study for the effects of organotoxins on plankton.

PCBs occur as dense, viscous, inert, clear liquids with varying degrees of chlorination. They are useful as electrical insulators, plasticizers, and lubricants. Although no longer in production, PCBs still in use enter the environment indirectly from such products as electrical transformers, lubricants, rubber, plastic, and paint, as well as directly from spills, such as those in the Duwamish River in 1974 and near Anacortes in 1980.   On entering seawater, PCBs sink directly to the bottom, mixing and dissolving little, and soak into the sediments.

Concentrations of dissolved PCBs in Puget Sound are highest in the industrialized waterways near the mouths of the Duwamish and Puyallup Rivers. Although there are high concentrations (up to 400 parts per trillion) in sewage effluent, the major PCB source for the Sound is the huge volume of river water entering the Whidbey basin, with a low level of PCBs from routine leakage and disposal. The major sinks, as for most pollutants in Puget Sound, are removal to the sediments and flushing out to sea.

The more such compounds are investigated, the greater the number found in the marine environment. A laundry list of potential hazards includes hexachlorobenzene, hexachlorobutadiene, pentachlorophenol, phthalates, and a variety of pesticides. Exceeding EPA standards in Seattle sewage effluent are, among others, chloroform, benzene, and pentachlorophenol.

Such organics are as toxic as any marine pollutant, showing deleterious effects on plankton at concentrations as low as one part per billion. The physiological modes of action of such chemicals on plankton are essentially unknown, and are likely to be as varied as their chemical structures. PCBs, for example, inhibit photosynthesis, but to different degrees in different phytoplankton. Recent evidence indicates that at least some PCBs can be broken down and detoxified in nature by bacteria or other causes.

Many such organic chemicals are poorly soluble in water. PCBs, in fact, are also heavier than water and tend to pool on the bottom. This insolubility means that PCBs have fates other than accumulation in seawater. One alternative path is to concentrate in the organic film that, because of surface tension, is found in the upper few millimeters of water. PCB concentrations in this layer have been estimated at five or more times those just below the surface, which can affect fish eggs and other specialized organisms, the neuston and periphyton, living at this interface.

Of all pollutants, organic chemicals have perhaps the strongest tendency to be absorbed by particles in water. Thus the insolubility of toxins, far from protecting plankton, actually makes plankton a major site of pollutant accumulation. As a result, the discussion of pollutant effects on plankton now merges with an examination of the role of plankton in the disposition of pollutants, and especially in the ways higher animals, including people, are affected.

Effects of Plankton on Pollution

In relatively clean waters, such as those of Puget Sound, plankters may actually have more influence on the fates of pollutants than pollutants do on plankton. When the two come in contact, the pollutants may be transposed, transformed, and transported by the plankton, and thus their effects on the rest of the marine food web may be altered or even controlled.

Absorption and Elimination

Figure 7.2 presents the concentrations of selected pollutants in Puget Sound plankton and suspended matter, mussels, and higher animals. When compared to the concentrations of dissolved pollutants (Figure 7.1), it is clear that these chemicals have an affinity for organisms, and selectively concentrate in them by a factor of a thousandfold or more.

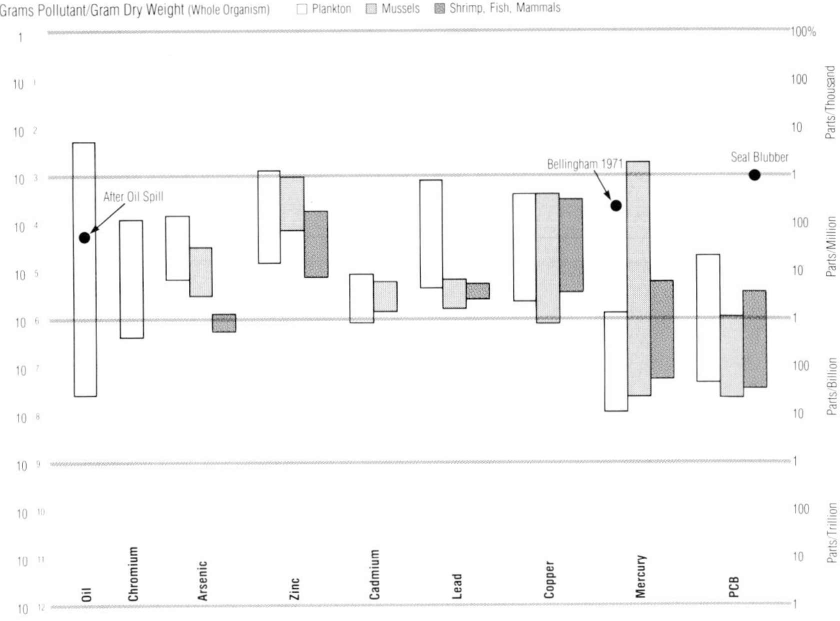

Figure 7.2 Concentration ranges (as a fraction of dry weight, assumed equal to one-sixth of wet weight) of pollutants measured in the tissues of Puget Sound plankton, mussels, and higher animals including fish, other shellfish, and mammals. Concentrations are roughly 1,000 times those in the surrounding water, demonstrating some bioaccumulation. Concentrations are similar at all trophic levels, however, providing little evidence of biomagnification.

This is a demonstration of bioaccumulation, the process by which organisms collect and store chemicals from their environment. The fact of bioaccumulation has been documented beyond any suspicion of analytical errors or experimental artifacts. The causes behind it, however, are both numerous and poorly understood, and data such as those in Figures 7.1 and 7.2 vary widely with season, organism, method of analysis, and location. One hundred parts per million of zinc, five times that of cleaner areas of the Sound, have been measured in plankton off West Point. Copper, at 90 ppm, is nearly 50 times higher in plankton off West Point than elsewhere. The highest levels of metal in plankton have been found at a dredge spoil disposal site in Elliott Bay, where lead at 886 ppm is nearly 200 times higher than in the cleaner plankton of Hood Canal.

Phytoplankters are, of course, specifically adapted for extracting scarce chemicals from seawater; nitrate and phosphate, for example, are present at levels of parts per million. Planktonic plants have evolved large surface areas and active metabolic pathways for taking up and storing nutrients, including such essential trace metals as zinc and copper, as well as such organic compounds as carbohydrates, amino acids, and vitamins. It should be no surprise, then, that toxic metals and organic compounds can be absorbed as well.

Accumulation may not be entirely biological because pollutants are equally concentrated in living and nonliving particles. Chemicals might simply be retained when organisms die and produce detritus, but there is evidence that living and dead particles accumulate at least some chemicals at comparable rates. Radioactive plutonium adheres equally rapidly to the surfaces of living and dead phytoplankton cells, and inert suspended matter from the Skagit River can absorb its own weight in oil. PCBs can transfer from nonliving to living particles. It may be, furthermore, that much of what appears to be uptake of metals into the protoplasm of plankters may simply be the adherence of tiny, colloidal metal particles to the outer surfaces of the organisms. The diatom *Ditylum*, one of the phytoplankters most sensitive to metal poisoning, secretes an outer mucous sheath, which it can slough off along with any pollutants adhering to it.

Part of the driving force for accumulation comes from the poor solubility of many metallic and organic compounds in seawater, and their corresponding affinity for organic matter (especially lipids) and even just inert surfaces. A gram of typical Puget Sound suspended matter, in fact, has a surface area of about 22 square meters. Crustaceans in general have an advantage over other animals because of the protection afforded them against direct absorption of pollutants from water by their exoskeleton, which in addition takes some of the body burden of toxins with it when the animals molt. But crustaceans are particularly sensitive to one class of pollutants, the organochloride insecticides, because of their close kinship with the insects those compounds are targeted to destroy.

The champion bioaccumulators are the filter-feeding bivalves, which even in the absence of their planktonic and suspended food will rapidly draw intense concentrations of pollutants across their gills and into their tissues from the large volumes of water they process. This ability, in fact, has contributed to the use of the common blue mussel, *Mytilus edulis*, as a worldwide early-warning system for marine pollution in a program called "Mussel Watch."

The concentrations of pollutants in exposed organisms and suspended matter do not increase indefinitely, but reach a plateau or saturation point. If placed in clean water, the pollutant burdens will decline. The bioaccumulation process is reversible to an extent; chemicals will migrate from zones of high concentration to low, whether into or out of organisms. Twenty-thousandfold accumulations of PCBs in phytoplankton can disappear after five days in clean water. Zooplankters use fecal pellets to rid themselves of toxins, perhaps even without assimilating them. PCB concentrations in fecal pellets of Mediterranean Sea euphausiids, for example, were 4 to 20 times those in the animals' phytoplankton food, several thousand times the levels in the animals' own bodies, and 1.5 million times those in the surrounding water. Toxins can remain permanently in some body tissues, however, particularly in fat and oil deposits or in nonliving skeletal components.

Transformation

Once associated with suspended matter, living or dead, pollutants do not necessarily retain their original identities. Some compounds, including petroleum hydrocarbons and even DDT and PCBs, are biodegradable. Many bacteria and plankters have the ability to alter or breakdown chemicals foreign to their systems.

Tin has been observed to undergo an organic transformation in plankton off California, which greatly increases its toxicity. Bacteria are implicated in such transformations as the conversion of the inorganic mercuric chloride emitted from the manufacturing facility on the shores of Minamata Bay, Japan, into organic methyl mercury, which has a thousandfold higher affinity for living tissue and a similar increase in toxicity. To that extent, those microorganisms shared the blame for the tragedy that resulted. The same process operates in the sediments of Bellingham Bay.

In contrast, however, bacteria and phytoplankton in Puget Sound convert inorganic arsenic compounds (which behave much like phosphate nutrients) to organic compounds, which are less toxic. There is also speculation that the unexplained liberation of dissolved organic compounds by phytoplankton may be a strategy to introduce chelators into the water, and so to tie up any potentially inhibitory metals.

More intriguing, however, are some of the mechanisms plants and animals use to cope with their internal body burdens of pollutants. Marine diatoms have been observed to break down DDT. Copepods from the vicinity of oil spills have been found to synthesize an enzyme (benzo (a) pyrene hydroxylase) that can dismantle one of the more toxic components of crude oil, and which is produced in response to the presence of oil. Phytoplankton and copepods contain a protein called metallothionein which can absorb and detoxify a certain quantity of metals. It appears, however, that these metallothionein-bound metals are not eliminated as quickly when organisms are placed again into clean water.

Transportation

The accumulation of pollutants in suspended matter, including plankton, merits special consideration because pollutant and particle thenceforth share the same physical fate. As the ultimate fate of most particles is to sink to the bottom (passing through various incarnations, living and dead), so many pollutants wind up buried in the sediments.

The extent to which suspended matter governs the fate of a particular pollutant depends on its relative affinities for water and particles. It is estimated, for example, that 40 to 90 percent of such poorly soluble pollutants as petroleum and mercury will come to rest in sediments, compared to less than 10 percent of the more soluble copper and cadmium. The removal of dissolved arsenic to the sediments is estimated to be 15 percent, mostly by adsorption onto clay.

Particles control the fate of poorly soluble pollutants despite bearing, at a given time, only a small fraction of the Sound's pollutant burden. Although pollutant concentrations in Puget Sound suspended matter may exceed those in water by a thousandfold or more, there are also at most nine parts of particles to a million parts of seawater. Nevertheless, the turnover of particles is so rapid that the constant replacement of this small fraction dominates other pollutant removal processes. Although it contains 20 percent of the PCBs in the Sound at a given time, for instance, sinking suspended matters removes to the sediments 80 percent of the Sound's dissolved PCB income. There is, in general, a good correlation between pollutant burden in suspended matter and in sediments at the same location.

The most important mechanism for delivering particles and the pollutants they contain to the bottom is the generation of fecal pellets by zooplankton. Copepods near a tanker accident off Nova Scotia were observed to ingest whole oil droplets without harm to themselves, eliminate them intact in their fecal pellets, and in so doing quickly deliver to the sediments 20 to 30 percent of all the oil spilled. Molted exoskeletons and dead carcasses also carry pollutants bottomward.

Thus, the importance of the transfer of pollutants from dissolved to particulate form is twofold. The accumulation of pollutants into the tissues of plankton, and the possible biomagnification at higher trophic levels, provide an avenue by which pollutants are channeled into fish and marine mammals to cause possible harm to them and to humans. But the greater effect of the same phenomenon may actually be a preventive one, from the point of view of the pelagic food chain—it may extract pollutants from the water and shunt them downward, out of the reach of pelagic animals, and into the sediments. Though a boon for the pelagic food chain, this bottomward diversion might have serious consequences for benthic animals, many of which are also important to the human economy. Much publicity has focused, for instance, on the health of bottomfishes in urban areas of the Sound.

Pollution and the Food Chain

Perhaps the greatest threat of pollutants to natural ecosystems, aquatic or terrestrial, is the potential for biomagnification of toxins from prey to successive predators, their effects worsening at each link of the food chain. The origin of the modern environmental movement can be traced to the discovery that birds, at the top of the food chain, suffered from pesticides directed at other organisms. Years of intense research since then have, as scientific inquiries often do, provided as many questions and exceptions as answers—especially in the marine environment, which behaves quite differently than the land. That biomagnification can occur under certain circumstances is not seriously disputed, but there is doubt and controversy about its importance relative to other pollution phenomena.

Pollutant concentrations in the tissues of animals at higher trophic levels in Puget Sound were presented in Figure 7.2. Biomagnification of mercury has been observed elsewhere in the large, predatory, and long-lived Atlantic swordfish, in the Pacific sperm whale, and in tuna. Museum specimens indicate that the former two species accumulated mercury long before humans began adding it to the environment. Mercury levels in most Puget Sound dogfish exceed the U.S. Food and Drug Administration standard of 0.5 parts per million for human consumption, so the catch must be exported. The high PCB levels in southern Sound harbor seals have the appearance of classic biomagnification, and have been tentatively linked to increased pup mortality in that location, as well as in southern California and in the Baltic Sea.

Several factors complicate the simple picture of increasing pollutant burdens at higher trophic levels caused by uptake from food. These include the exchange of pollutants directly with the water (or sediment), the abilities of organisms to transform and eliminate pollutants, and the peculiarities of individual chemical and biological species. Magnification patterns are also complicated by the varied diets, life cycles, and migratory patterns of plankton and higher animals.

Many field tests of biomagnification have met with mixed results. As evident from Figure 7.2, pollutant levels in some higher organisms in Puget Sound, including such planktivores as fish and shrimp, can be lower than those in plankton, which ostensibly occupies lower trophic levels. In such comparisons the artificial concepts of "food chain" and "food web" begin to lose their utility, for in the maze of dietary connections in Puget Sound, trophic levels blur. Attempts to test for biomagnification by tracing increases in nonpollutant elements (specifically, the ratio of cesium to potassium) at higher trophic levels in a food web off California have proven inconclusive as well. The highly publicized fish diseases in urban areas of the Sound can be caused by direct contact with contaminated sediment, and so do not necessarily result from biomagnification at all.

The occurrence of biomagnification, and its importance relative to other modes of contamination of organisms, seems to depend on the relationships of various types of pollutants and organisms to water. Pollutants that are poorly soluble in water (hydrophobic) and more soluble in fat (lipophilic)—including petroleum, chlorinated hydrocarbons, and mercury—are more likely to be bioaccumulated in living tissue, as well as in detritus and organic sediments. They are partitioned from water and retained in lipid tissues, and are difficult to excrete unless chemically transformed. Such pollutants tend to accumulate most in organisms that live the longest, and which have a higher fat content, such as dogfish and harbor seals. Such pollutants likewise have a higher potential for transfer up the food chain, and thus for biomagnification.

There are also differences between animals based on the degree of direct exposure of their tissues to water. Gilled animals, especially those that pump water to feed, are likely to have higher rates of pollutant exchange (both uptake from and elimination into ambient water) than animals with impermeable body surfaces. Thus rates of pollutant exchange would be higher in suspension-feeding bivalves than in carnivorous fishes, and higher in either of those than in air-breathing marine birds and mammals. A faster rate of exchange implies that observed body burdens of pollutants result from simple bioaccumulation from water, rather than from biomagnification. Animals at the tops of marine food chains, especially birds and mammals, exhibit the greatest biomagnification because of their long life spans, their high fat contents, and their reduced ability to exchange acquired toxins with the water, as well as because of their higher trophic status.

The differential effects of pollutants on various species have an important manifestation in Puget Sound plankton, as revealed by studies in neighboring waters. A major study in Saanich Inlet, B.C., tested the effects of addition of oil, PCBs, copper, and mercury to large plastic cylinders of seawater. The study was called CEPEX, for Controlled Ecosystem Pollution Experiment. The CEPEX study found that these pollutants did more than simply reduce the standing stocks of plankters or inject toxins into the food chain. Pollutants selectively crippled the diatom-based primary food chain, that collection of plankters thought to favor the growth of salmon and other valuable pelagic fishes.

There is evidence from elsewhere in the sea, as well as from CEPEX, that large, centric diatoms are the most sensitive phytoplankters to all kinds of pollutant stress. In studies from Long Island Sound, New York, the diatoms *Skeletonema*, *Thalassiosira*, and *Chaetoceros* suffered reduced growth at PCB concentrations as low as one part per billion, while the pennate diatom *Nitzschia* and the green flagellate *Dunaliella* were unaffected by levels of up to 100 parts per billion. Centric diatoms are sensitive to concentrations of 50 parts per billion of DDT, while *Dunaliella* is resistant to one part per million. Similar results were found with the insecticides Chlordane and Dieldrin, with the latter selectively eliminating phytoplankters larger than a certain size. Laboratory experiments parallelling CEPEX showed that low levels of petroleum hydrocarbons (50 parts per billion) selectively stimulated the growth of phytoflagellates and small pennate diatoms, the same groups that populate Saanich Inlet during off-bloom periods. Zooplankton groups may also differ in their sensitivities to pollutants, but such effects are difficult to distinguish from effects of altered phytoplankton diet.

The most conclusive results, however, came from CEPEX itself. When copper was added at 50 parts per billion, large centric diatoms were replaced by an equivalent biomass of phytoflagellates and small pennate diatoms, the base of the secondary branch of the food web. This replacement happened in unpolluted enclosures, but the effects were more pronounced under copper stress. The surviving organisms demonstrated a tolerance to high levels of copper, compared to untreated organisms of the same species. The addition of copper also stimulated a rapid increase in the release of organic carbon by phytoplankton, and a subsequent explosion in bacterial populations, supporting the suspicion that such "excretion" may be a deliberate behavior to reduce metal toxicity by chelation. The small phytoplankters could no longer be harvested by the large zooplankton present, which in addition suffered some direct toxicity from the copper, and so died off.

Different results were obtained when one and five parts per billion of methylmercury were added to enclosures containing fresh experimental populations. An initial period of drastic decline in phytoplankton populations was followed by a recovery, with some demonstration of an acquired tolerance to mercury. The population to which one part per billion had been added grew back as the predicted small flagellates, but large diatoms and dinoflagellates dominated the recovery in the more polluted enclosure, and productivity after two months exceeded that in the unpolluted enclosure—hardly what would be expected from severe pollution.

At five parts per billion, mercury appeared to affect the zooplankton more than the phytoplankton. When mercury was added, the copepods *Calanus* and *Pseudocalanus* and their larvae disappeared, leaving the larvacean *Oikopleura* as the dominant zooplankter. With the small phytoplankton removed by *Oikopleura* and the copepods removed by mercury, the large diatoms and dinoflagellates were free to bloom without interference. The copepods, furthermore, were never able to recover because their nauplii could not survive without small phytoplankton. Finally, juvenile salmon inhabiting the enclosure, while not directly affected by the mercury, starved for lack of copepods to eat.

These results demonstrate the complexity of pollutant effects, and illustrate indirect damage to animals by elimination of their food supplies when no direct toxicity is present.

The conclusion distilled from all this is that the effects of pollutants on the planktonic food chain, and on the higher animals that depend on that food chain, cannot be interpreted in any simplistic fashion. There are direct toxic effects of pollutants on both plants and animals. Biomagnification can occur as toxins pass from prey to predator, but serious consequences seem limited to particular pollutants and specific animals. When overall primary productivity is reduced as a result of pollution, animals can be affected by reduced food (and perhaps oxygen) supply. But changes in the quality of food supply can be as catastrophic as reductions in quantity; animals cannot exploit an improper food, however abundant that food may be. The most significant consequence of polluting the plankton might be a shift away from the normal population balance, toward some unknown new community, with unpredictable results on the food web.

CHAPTER EIGHT

Red Tides

> About, about, in reel and rout
> The death fires danced at night
> The water, like a witch's oils,
> Burnt green and blue and white.
>
> Samuel Taylor Coleridge, *The Rime of the Ancient Mariner*

Perhaps the manifestations of plankton best known to Puget Sound residents are phenomena called "colored water"—red, brown, and luminescent—collectively known as "red tide." Red tides occur anytime from spring through fall, though they are most common during late summer. They form impressive displays in bays, along shorelines, and in boat wakes. Most red tides are caused by one of several genera of pigmented dinoflagellates, and most are nontoxic. *Gonyaulax, Gymnodinium, Ceratium,* and *Prorocentrum* appear to be toxic; but *Peridinium,* the animal *Noctiluca,* and the unique symbiotic protozoan *Mesodinium* are nontoxic. Many of these organisms are also bioluminescent, as are some crustaceans, ctenophores, medusae, larvaceans, annelid worms, and fishes.

Any of these organisms can form a "bloom" dense enough to discolor water at the surface, particularly when acted on by physical concentrating forces. They are often seen in windrows or discrete patches, where they accumulate like flotsam. *Gymnodimium* colors the water a deep pink or chocolate brown, *Noctiluca* a tomato-soup red, and *Mesodinium* brick red to purple. The colors come from accessory pigments to the ever present chlorophyll, which are related to the pigments we see unmasked in autumn leaves.

However pleasing to the eye, these incidences can sometimes have sinister effects, resulting in morbidity and mortality of both marine animals and humans. The two most prominent effects are paralytic shellfish poisoning and oyster larvae mortality.

Paralytic Shellfish Poisoning (PSP)

In June 1793, Captain George Vancouver and his crew were reconnoitering the area around the central coast of British Columbia when one party of four men pried some mussels from the rocks of a small cove for their breakfast, as they had done on many previous mornings. On this unfortunate day, however, all four of the men, within a few minutes after breakfasting, were beset by a numbness of the lips and fingertips. This was followed by paralysis of the arms and legs, a feeling of dizziness, and nausea. For one man, John Carter, it meant death. The

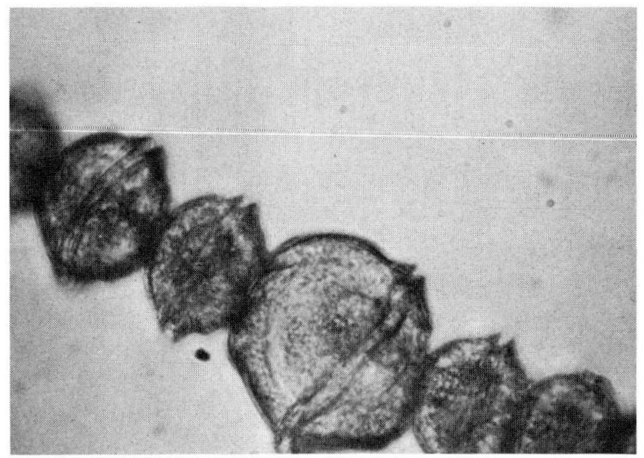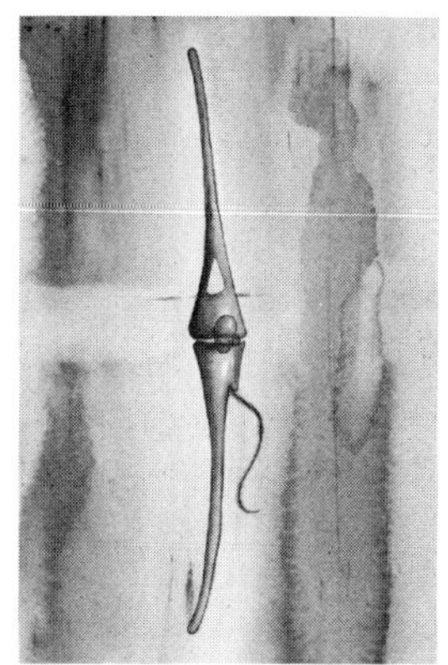

Figure 8.1 *Gonyaulax (Protogonyaulax) catenella* (above left), the chain-forming dinoflagellate responsible for paralytic shellfish poisoning in Puget Sound. The enlarged cell (center) has been infected by the parasitic dinoflagellate *Amoebophyra ceratii*, and will soon burst and die. Actual size approximately 30 micrometers per cell. (Photo courtesy Louisa Nishitani). *Ceratium fusus* (above right), a spiny dinoflagellate linked to oyster larvae mortality in the southern Sound. Actual length approximately 100 micrometers. (After Cardwell et al., 1977)

other three men, acting on the advice of a fellow crewman who had experienced a similar malady in England, exercised vigorously and, coincidentally, survived. They named the location Poison Cove. This was the first recorded incidence of paralytic shellfish poisoning (PSP) on the Pacific coast of North America.

Paralytic shellfish poisoning is found on coasts throughout the world, including Northern Europe, Southern Africa, Japan, and both coasts of North America. God is alleged to have slain Egyptians by this means on Moses' behalf: "All the water changed into blood. The fish died and the river stank" (Exodus 7:20-21). It was not until an outbreak of shellfish poisoning in California in 1927 killed several persons and sickened over one hundred more that the poison was connected to red tides. One scientist, thinking that it might have come from the food of the offending mussels, managed to isolate a species of phytoplankton that produced a toxin matching that in the shellfish. This was the first instance in which plankton was found to be an enemy, albeit an unwitting one, of humans.

Although there have been repeated outbreaks in the Puget Sound area, notably with several deaths in 1942 and 1957 in Canada and Washington, it was not until an outbreak in the northern Strait of Georgia in 1965 that the local planktonic culprit was first identified. It is now thought to be *Gonyaulax* (or *Protogonyaulax*) *catenella*, an armored dinoflagellate that ranges from 15 to 55 microns in diameter, and which unlike most dinoflagellates occurs in chains of from two to eight cells (Figure 8.1). It has been grown in culture and its life habits studied. *G. catanella* is ten times more toxic in Washington than in California, but only a tenth as toxic as a sibling species (variously called *G. tamarensis* or *G. excavata*) that has been responsible for poisoning and death from Long Island Sound to Nova Scotia.

Although sparse *G. catenella* populations had been detected before, the first PSP related health problems in the inner waters of Puget Sound appeared in September of 1978, during warm "Indian Summer" weather following heavy rains in August and September. They occurred on Saratoga Passage along the eastern shore of Whidbey Island, and gained public notice when nine people became sick, four of them seriously enough to be hospitalized. PSP spread southward as far as Vashon Island over the course of three weeks, necessitating closure of beaches to shellfish gathering. Toxin levels of 30,000 micrograms per 100 grams of tissue (80 is sufficient to close a beach) were found in shellfish meat. The PSP did not reach south of The Narrows, nor into Hood Canal, and by the end of the month some beaches were reopened in Puget Sound. Beaches on Whidbey Island and the San Juans never reopened, however, and additional closures began again the following April. By July 1979, all beaches north of Tacoma were again closed. (A fatality occurred in the northern Strait of Georgia in May 1980.) It is still a mystery why PSP-related illnesses were never reported from within Puget Sound proper until 1978.

Of all the animals in Puget Sound that feed on phytoplankton, primarily only bivalves acquire PSP and are eaten by humans: the mussels *Mytilus edulis* and *M. californianus*, the oysters *Crassostrea* and *Ostrea*; the butter clam *Saxidomus*, the soft-shell clam *Mya*, the Manila clam *Venerupis*, the littleneck clam *Protothaca*, the cockle *Clinocardium*, and some others. All of these bivalves obtain their food by pumping water into their shells and over their gills, where plant cells are trapped and directed toward the mouth. The mussels and oysters, which live on rocks and atop the sediment, simply open their shells to obtain water. The clams, living below the sediment, have siphons that extend into the overlying water. All of these bivalves live in the intertidal zone near the water's surface, where phytoplankton is most abundant. Mussels can become poisonous within a few days of an outbreak of *Gonyaulax*, making them an excellent organism for monitoring the onset of toxicity. Mussels also rid themselves of toxin soon after a bloom. In contrast, the butter clam, which concentrates the toxin in its siphon, can retain it there for as long as two years after the initial outbreak.

The effects of PSP are of course worst when eaten on an empty stomach. There is no antidote; treatment is to induce vomiting as soon as possible, and to administer a fast-acting laxative to minimize the absorption of the toxins. In the most serious cases, death occurs by respiratory paralysis within 2 to 12 hours. The principal (but not only) toxin

Toxin	Dose
Botulinus	0.00003
Tetanus	0.001
Diptheria	0.3
Cobra Venom	0.3
*Saxitoxin	9.0
Curare	500
Strychnine	500
Muscarin, from *Amanita* mushroom	1100
Sodium Cyanide	10,000

produced by *Gonyaulax*, saxitoxin, has been isolated and synthesized artificially. Saxitoxin is an alkaloid nerve poison (neurotoxin) that acts by blocking transmission of impulses from nerve to muscle, and is one of the more potent natural poisons (Table 8.1). Pure samples of it are kept on hand by the Army at Fort Dietrich, Maryland, in case the Russians land looking for clams.

There is no truth to beliefs that garlic will neutralize PSP toxins, or that they can be detected by the discoloration of silver in cooking water. High-temperature frying can reduce the toxicity slightly, and a portion of it can also be discarded with steaming liquor. Nevertheless, enough toxicity can remain to sicken or kill humans. A dangerous old-wive's tale claimed that shellfish were safe to eat in the months of the year which contained the letter "R"—that is, that they were unsafe from May through August. Blooms of *Gonyaulax* can occur from April to November, however, and shellfish can be toxic all year. The only assurance of safety is the method used by the Health Department for assaying the toxins in shellfish tissues. A standardized amount of meat is ground up and injected into a white mouse. If nothing happens, the stuff is safe to eat. If the mouse suddenly stiffens, begins wildly leaping about its quarters, gasps for breath and dies, so might you. The length of time it takes for the mouse to die is a measure of the strength of the toxins.

PSP toxins do not seem to harm most shellfish in which they are concentrated, despite their toxicity to humans and other, mostly warm-blooded animals. Cats in particular are extremely vulnerable. The reasons for these sensitivity differences are still not well understood, but the toxins are apparently harmless unless acted upon by acid, as occurs during digestion by humans but seemingly not by shellfish. When acid-hydrolyzed saxitoxin is injected into many fish and shellfish, they show clear signs of poisoning; but conversely, some shellfish have an apparent ability to break down the toxin. On the Atlantic Coast, large kills of herring have been blamed on their ingestion of zooplankton which had eaten *Gonyaulax*, and further kills have been traced to other red tide organisms. PSP toxins are found in some other animals, but the place of *Gonyaulax* in the food web has not yet been determined.

Several competing theories still have not fully explained the appearances of PSP. *Gonyaulax* seems to like quite warm water, around 14°C (a typical surface temperature in the main basin in summer), and can grow well at any salinity encountered in Puget Sound. *Gonyaulax* seems to appear most frequently in protected inlets during later summer spells of prolonged warm clear weather, when nutrients are depleted from a stratified surface. Its summer abundance is inverse to that of diatoms. Before invading Puget Sound, it was historically most common in Sequim Bay (on the southern shore of the Strait of Juan de Fuca), which is almost completely isolated from the Strait by a narrow mouth and a sandspit, and receives little freshwater runoff. Its abundance there has been correlated with both the intensity of sunshine and the duration of spells of clear weather. Mixing is quite slow here during the summer, and experiments in New England demonstrated that just stirring a flask of different species of *Gonyaulax* (*G. excavata*) stopped its growth.

Gonyaulax also appears during periods in which rainfall (particularly a large amount of rainfall without heavy winds to cause mixing) is followed by prolonged clear weather. This may produce necessary stratification. Laboratory cultures of *Gonyaulax* can be stimulated by extracts of terrestrial soil, however; an essential nutrient may thus be provided by the runoff from rainfall, or the organic chemicals in soil may chelate toxic trace metals. More speculative yet is the hypothesis that in some parts of the world PSP may be getting worse due to the chelating effect of sewage dumped into the sea.

The appearances of PSP, like those of phytoplankton in general, are determined in part by localized circulation patterns in coastal and estuarine areas. Although the results have not yet been applied to Puget Sound, examples from England, Florida, and Maine show that dinoflagellates (including *Gonyaulax*) accumulate where highly stratified waters meet mixed waters. Invisible blooms can occur below the surface in stratified, nutrient depleted areas, and blooms also can be highly patchy. This historically has been a problem in observing shellfish toxicity on the long, sparsely populated shoreline of British Columbia where a toxic bloom could appear, last a few days, and depart before scientists could even know it existed—much less study it—and where shellfish taken only a mile apart could vary twentyfold in toxicity.

Finding PSP outbreaks in time to warn the public is complicated by two more problems. Shellfish can acquire toxicity even when *Gonyaulax* is too dilute to color the water. Furthermore, *Gonyaulax*, like other neritic phytoplankters, forms resting cysts under certain conditions, which fare yet to be determined. These cysts can be more toxic than the swimming cells, and can contaminate shellfish throughout the year. Once cysts are dropped to the bottom of a body of water, there apparently is no way to eliminate them, although they eventually will die if they do not germinate. The cysts may be spread to uncontaminated areas by dredging and by transportation of shellfish that contain

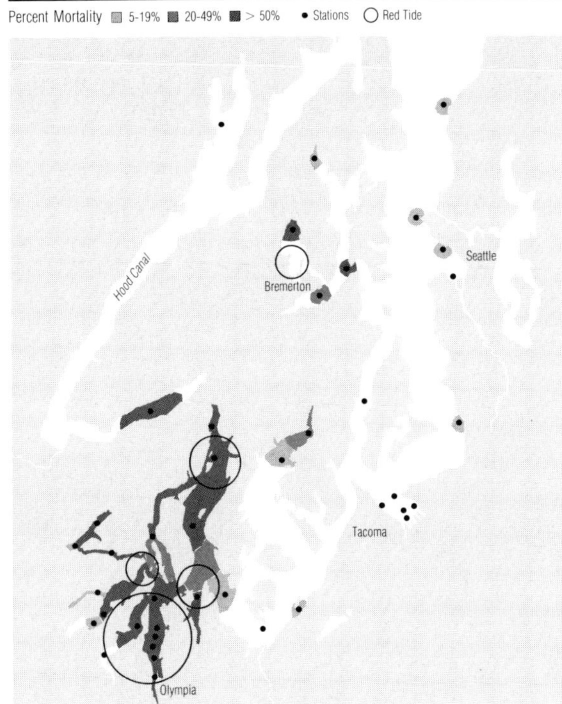

Figure 8.2 Oyster larvae mortalities in the central and southern Sound, as determined by surface water bioassays during 1977. Visible red tides are noted. (After Cardwell et al., 1979)

them. Low levels of PSP that appeared in the southern Sound in 1981 may have originated in this manner.

A couple of promising developments make the story of *Gonyaulax* not an entirely negative one, however. Saxitoxin shows some promise as an insecticide. Research at the University of Washington has also uncovered a dinoflagellate, *Amoebophyra ceratii*, that can parasitize several other dinoflagellates—including *Gonyaulax*—in the wild, and may in fact exert some control on outbreaks of PSP. This discovery suggests the possibility that *Gonyaulax catenella* might be controlled in confined bays by growing *Amoebophyra* on a nontoxic host such as *Peridinium* in the laboratory, and releasing it in those bays.

Other Toxic Red Tides

Although southern Puget Sound may (so far) have fewer problems with paralytic shellfish poisoning than the northern Sound, recent research has disclosed a red tide-related problem of a different sort. Very high levels of oyster larvae mortality—over 50 percent of the larvae tested in bioassays—have frequently been observed in shallow inlets of the southern, central, and northern Sound, and Hood Canal (Figure 8.2). In some cases these have been correlated with pulp mill effluent (e.g., Port Gardner, Bellingham Bay), but in the southern Sound some peculiar traits emerge. Well-mixed areas, even those very close to areas of high mortality, are free of the problem. The toxicity is often most severe at a depth of roughly three meters. Larvae are not sickened by small amounts of the harmful water (as they might be if a chemical pollutant were responsible), but rather are either totally unaffected or are killed.

Oyster larvae mortality correlates best, it turns out, with abundances of the dinoflagellates *Ceratium fusus* (Figure 8.1) and *Gymnodinium splendens*, and sometimes *Prorocentrum gracile*. They seem not to be poisonous, as *Gonyaulax* is; the water by itself is nontoxic, either without cells or with ground-up cells. Instead, *Ceratium* at least may impale the larvae on its long spines.

Similar effects are observed when larvae are exposed to sharp-edged particles from dredge spoils. A kill at a commercial oyster hatchery on Liberty Bay in 1975, first thought to have been caused by mercury from a nearby Navy installation, was later attributed to a red tide. Larval bioassay results also correspond closely with observed mortality of adult oysters in the southern Sound. *Ceratium* blooms were observed years ago in such places as Hood Canal and San Juan channel, before they were linked to mortalities.

Toxins have recently been isolated from related phytoplankton species, including *Gymnodium breve*, which causes fish kills in Florida, and *Prorocentrum minimum* from the Strait of Georgia. There also appears to be a link between the phytoflagellate *Olisthodiscus luteus* and fish mortality in northern Puget Sound and elsewhere, seemingly due to bacterial stimulation by carbon excreted from the alga. The dinoflagellate *Gyrodinium* has been linked to kills of shoreline fish, snails, and limpets in Ireland, and large kills of shellfish have been caused off New York City by oxygen depletion from decay of red tide blooms of another species of *Ceratium* (*C. tripos*).

Are all these harmful red tides really increasing in frequency and severity, as seems to be the case, and if so, why? That question cannot yet be positively answered. The increases in red tides recorded within the last few decades do appear to be real, but they may stem from increased surveillance and reporting, due to higher population density and better scientific knowledge. There may be long-term fluctuations in red tide incidence related to natural causes such as climatic cycles, of which we are seeing only a small segment. The possible link between human activities and red tides is still a speculative one. Every theory of the effect of pollution on red tides has its counter-theory, and as yet none has been reliably confirmed. Certainly toxic red tides have occurred and still cause problems where they could not have been aggravated by civilization nor scrutinized by science. At worst, humans have perhaps only exacerbated an entirely normal process of nature.

CHAPTER NINE

Marine Pastures

There, said he, pointing to the sea, is a green pasture where our children's
grandchildren will go for bread.

Obed Macy, *A History of Nantucket*

Perhaps the most persistent belief about life in the sea is that one
day food from the sea will eliminate world hunger. This myth has been
perpetuated by people with more idealism than knowledge, for the re-
alization of this worthy and far-sighted goal is at best years or decades
away. There are substantial obstacles to developing the ocean's food
supply—obstacles for which there would seem to be no easy solutions.

The first problem is in finding any more food to harvest. Certain of
the world's fish stocks appear already to be badly overfished, and the
area within which to pursue new fisheries is limited to the more productive coastal and polar regions. Although global phytoplankton production exceeds the global harvest of cultivated crops by a factor of
five, most of the world's plankton production takes place in the open
seas, where it is so widely dispersed that it is impractical either to harvest it directly, or even to harvest the fishes it supports. The higher
rates of plankton and fish productivity, the technology of harvesting,
and even issues of international law all favor a concentration of effort
in the areas already exploited.

The strong stratification responsible for this low productivity in
mid-ocean also restricts the potential for artificially enhancing the productivity of the sea at large. Some means would be needed to stir the
ocean to depths of as much as 100 meters over a wide area, in order to
bring up enough nutrients to start the food chain going, and even then it
is not clear what sorts of organisms would grow. Perhaps the energy for
this pumping can someday be derived from projects on the drawing
boards today, such as Ocean Thermal Energy Conversion (a sort of floating reverse heat pump, which would generate electricity by pumping
deep water to the surface), or from wave or tide power. Clearly, global
cultivation of the ocean would require forces not yet even imagined.

The barren condition of the open ocean makes coastal areas such as
Puget Sound very important to the future harvest of the sea. Here, many
technical difficulties are reduced—the area is easily accessible, well-
mixed, and already quite productive. Schemes for extracting more seafood from coastal areas fall into the same categories as food-providing
schemes on land: hunting, ranching, and farming. All have applications to the role of plankton in the harvest of food from Puget Sound.

Hunting

Marine hunters, the commercial and sport fishermen, harvest seafood without participating in its rearing. At the present time, increasing the hunting effort is the quickest and cheapest method of increasing the yield from the sea. New fishing equipment is more efficient at catching those fishes and shellfish that are already being harvested, and with the advent of the 200-mile offshore fishing limit there are new possibilities for exploiting previously underutilized species.

The ceiling on natural fish catch is imposed ultimately by the production of plankton. The types of plankton that grow also determine (and are determined by) the species of fish, shellfish, and other animals that grow. Coastal waters are believed to be most productive because they are well-mixed and because they have large plankters and short food chains with a minimum of tropic inefficiency. When harvesting fish populations, both the natural limits on production and the complex interactions that account for production must be considered.

Peru is a classic example of how attempts to hunt more food from the sea can go awry. Coastal upwelling enriches the coast of Peru and supports a vigorous growth of large diatoms, fed upon by anchovies—a highly-efficient, two-link food chain. At one time, one in every five pounds of fish caught in the world was caught off Peru. But in the waters off South America, as in Puget Sound, biological productivity is intimately tied to the physics of the ocean, and to the weather. Periodically, upwelling ceases, nutrients become scarce, and the planktonic food supply for the anchovies is cut drastically, apparently because of large-scale weather patterns over the South Pacific. The large diatoms are frequently replaced by flagellates, or even by red tides.

Such an event is called an *El Niño* (Spanish for "The Christ Child," since it appears most frequently around Christmas), and has apparently always been a normal feature of the region's oceanography. The anchovies dependent on the diatom production simply die back during an *El Niño*, then rebuild themselves when favorable conditions return. In the early 1970s, however, unregulated numbers of fishermen were already taking almost as much fish tonnage as the system would sustain (or perhaps more), when a sequence of *El Niños* decimated the stock. Desperate fishermen worked all the harder to catch what they could. The unfortunate result was that the anchovy population dropped to a level from which it may never recover.

A similar fate may also be overtaking North Pacific and Puget Sound fisheries. The potential harvest of the sea is limited both by the area of the sea from which we can expect to harvest fish, and by a ceiling on the harvestable fish within that area. Both of these limits are set by the plankton and by the meteorological and physical processes that govern plankton production.

One proposed escape from catch limitations is to harvest organisms lower on the food chain; that is, to harvest plankton directly. Although the idea is not totally new—jellyfishes are a traditional food in Asia—its magnitude is new. Already, the richest crop of plankton in the world, the large herbivorous antarctic krill, *Euphausia superba*, is the subject of trial harvests by the Russians and Japanese, and many more countries are contemplating joining them. Antarctic krill grows rapidly and has a standing crop of 40 to 1,000 million metric tons (44 to 1,100 million long tons), compared to the annual world fish harvest of about 60 metric million tons (66 million long tons). Krill has the flavor and nutritional value of its larger cousin, the shrimp, although it is said to resemble cooked maggots in appearance. Antarctic waters are currently unregulated and are open to international exploitation. Additionally, because of the overharvesting of one of the krill's principal predators, the blue whale, there may be an uneaten "surplus" of many millions of tons of krill.

There has been speculation about harvesting copepods, although these smaller crustaceans are more difficult to collect and contain a higher proportion of indigestible chitin exoskeleton. Zooplankton could be used both for animal and human food, as either a dish unto itself or ground into meal as a protein supplement.

There are problems with this potential harvest, however. It is not clear, first of all, that a surplus of antarctic krill really exists, and it is difficult to determine what effect removal of large quantities of krill could have on the remaining whale populations, or on seals and birds. Some preliminary euphausiid harvests have been conducted in the Strait of Georgia for a number of years, but an extensive zooplankton fishery in Puget Sound and the North Pacific would almost certainly conflict with the fishery supported by a major zooplankton predator—the popular, tasty, high-priced Pacific salmon.

The main obstacles to the exploitation of plankton as a food source are technical and economical. Put simply, the question is whether enough plankton can be caught, using state-of-the-art plankton nets, to pay for the effort of pursuing it. Although seafoods in general require less energy than terrestrial foods to bring to market, the plankton resource is so dilute, and the organisms so tiny, that it seems more economical in most cases to allow specialized animals to harvest for us and to accept the loss of efficiency in the food chain than to try to beat nekton at its own game. Imagine, for comparison, humans gathering nectar to make honey, instead of allowing bees to do it for them. The prospects for harvesting zooplankton are supported only because it is found in dense swarms—patches—detectable with echo sounders. Ultimately, the feasibility of zooplankton fisheries will depend on the price it will bring at the grocery store.

One of the most successful and biologically efficient methods of obtaining a harvest from the plankton has been to gather naturally occurring planktivorous shellfish, such as clams, oysters, and mussels. These animals are efficient and powerful at extracting plankton and suspended matter from moving water, and have few predators that compete with humans. Potential shellfish harvest in the wild is even more limited than that of finfishes, however, since the area of suitably productive shallow water habitat is less than the area of open coastal waters. Furthermore, natural growth of some shellfish species is hampered by sporadic failures in larval production and high mortality due to cold water, predators, and pollution. New mechanical shellfish harvesting technologies, such as water jets, conveyor belts, and suction dredges, promise increased supplies and reduced costs for this mode of marine hunting, but are running afoul of political and environmental complications. To increase the harvest of seafood, therefore, scientists and growers are turning to ranching and farming techniques.

Ranching

Marine ranchers, like their terrestrial counterparts, breed their stock, rear them as young, then turn them loose to feed and mature on open rangeland before finally rounding them up for sale or slaughter. Puget Sound ranchers raise fish and shellfish in hatcheries, which reduce reproductive mortality by providing suitable spawning and rearing habitat. It is a form of ranching because the animals are released from human control and allowed to graze freely in their marine pastures, then are harvested upon their return, and because there is a degree of selective breeding.

Ranching techniques have achieved their greatest successes with bivalve molluscs. For years clam and oyster larvae have been reared under controlled laboratory conditions, using a phytoplankton diet, until they could be transplanted to a beach as spat. The rancher may thus breed the shellfish stocks selectively, feed the larvae an optimal food at an optimal temperature and salinity, add antibiotics and vitamins, and control the time and place at which the spat are introduced into the wild, selecting favorable environmental conditions and reducing competition for food and space. Washington growers once imported most of their spat for the Pacific oyster from Japan, since natural sets in Puget Sound are not sufficiently reliable or predictable for commercial operations. Local hatcheries now produce most of the spat planted in Puget Sound oyster beds, however. In addition, the Washington Department of Fisheries has recently begun raising razor clams (*Siliqua patula*) for planting on outer coast beaches, and experiments are underway elsewhere for the raising of Dungeness crab (*Cancer magister*), geoducks (*Panope generosa*), and abalone (*Haliotis* sp.).

Hatcheries have also been successful in maintaining the salmon harvest. In terms of yield per unit of area required and feed consumed, and therefore per dollar spent, such hatcheries are the best investment available for increasing the current yield of the sea. This method cannot increase the capacity of marine ecosystems for producing animals, but in places like the Pacific Northwest, where unexploited plankton food may be available because of declining of salmon stocks, and where many natural spawning grounds have been destroyed, hatcheries can help maintain viable fish populations. Still, the potential benefits of hatcheries are constrained by other limits on salmon populations, including overharvest and loss of genetic diversity. Hatchery rearing of salmon has been carried out in the Puget Sound area for decades, and expansion of salmon-rearing efforts remains one of the best hopes for maintaining the yield of protein from Northwest inland waters.

Farming

Marine farming, more popularly known as aquaculture or mariculture, entails the husbandry of marine animals (and some plants) from birth to death. Aquaculture is amazingly undeveloped in the United States considering (or perhaps because of) our technological advancement. Americans obtain less than 10 percent of their protein from aquatic animals, eating only 5.5 kilograms (12 pounds) of fish per capita per year, compared to 32 kilograms (70 pounds) in Japan. American aquaculture production is minimal compared to that of the Orient and satisfies but a small portion of our seafood appetite. Research on marine farming falls into two categories: raising planktivorous animals on or in a confined structure within a natural body of water, and growing plankton and planktivores artificially in completely enclosed systems.

One form of mariculture emerging on Puget Sound is net-pen rearing of salmon and cutthroat trout. The fish are kept in large pens immersed in Puget Sound from the time they leave the hatchery as juveniles until they reach a marketable weight of about one pound, called "pan-size." The advantages of this approach are that animals are not lost to either migration or predation, and that they live in very nearly their natural environment, saving growers the trouble and expense of duplicating those conditions artificially. Similar experiments aimed at raising prawns have met with more difficulty, owing to a more complex life cycle.

The disadvantage of net-pen farming is that although natural zooplankton is available, fish populations are so dense that their diet must be augmented with costly feed. Fish cultivation in general is ecologically more efficient than red meat production, but supplementing salmon diets amounts simply to converting one expensive food into another, with a net loss of energy. Also, as in hatcheries, crowding and genetic uniformity resulting from controlled breeding make fish more susceptible to oxygen depletion and disease. Finally, some forms of plankton can even be hazardous to the penned fish; dense phytoplankton blooms of spiny *Chaetoceros* and *Ceratium* have been known to puncture the sensitive gills of young fish and cause heavy mortality because the animals are not free to swim away from the algal masses as they would in the wild. Nevertheless, because of the attractive prices and demand for salmon, net-pen farming is achieving commercial success on Puget Sound.

Currently the most successful form of marine farming worldwide— second only to hatcheries in efficient use of energy and space, and second only to seaweeds in global tonnage of cultivated harvest—is the raising of bivalves using suspension culture. Oysters and mussels, in particular, are allowed to attach to ropes or are placed in cages as spat, then hung from posts, buoys, or rafts in waters that offer both suitable protection from storms and healthy flushing by currents, tides, and winds. Suspending bivalves off the bottom tremendously increases the volume of water from which these animals can filter their phytoplankton food by extending their habitat both vertically and offshore. Suspension culture maintains more animals in the productive surface layer and eliminates exposure during low tides. Shellfish in suspension culture reach market size twice as fast as those in bottom culture and produce a higher quality of meat since less bottom sediment is ingested.

One advantage of using shellfish as a means of extracting more plankton protein from the sea is that the shellfish are extremely efficient filterers and are able to capture very small plant prey such as flagellates. This is a simple, two-link food chain, with a minimum of the trophic inefficiency that accompanies longer food chains, and the crop is sessile and relatively easy to harvest. The yield of raft cultures of mussels approaches the phenomenal figure of 30,000 metric tons (33,000 long tons, fresh weight) of meat per square kilometer of raft area per year. That compares to yields of a few hundred to a few thousand metric tons for bottom culture of oysters; less than a hundred metric tons for coastal finfisheries such as those of Puget Sound, Georges Bank, and the North Sea; and at most 50 metric tons of beef or pork on a feedlot. These figures further demonstrate that the potential productivity of water is much higher than that of land—if the limitations of fluidity can be overcome—because marine bivalves do not have to expend as much energy as terrestrial animals in supporting themselves, moving about, or maintaining their body temperatures, and so can devote proportionally more energy to protein production.

Suspension culture of oysters is beginning in Washington State, and there is potential for mussel culture as well. Some problems still must be dealt with in raft culture: red tides, bacteria, viruses, and pollutants can contaminate entire crops. Labor is expensive, and on Puget Sound there is an aesthetic consideration as well: it can be difficult to convince citizens to accept rafts, buoys, and posts along their expensive waterfronts and in their favorite boating spots.

The ultimate marine farm would not depend on natural plankton production. Artificial stimulation of phytoplankton production in a natural setting to enhance the yield of shellfish farming has been successful in ponds in Alaska and in the Virgin Islands. These experiments involved pumping deep, nutrient-rich water to the surface, in an enclosed area where enrichment effects would not be dissipated, and where raft or net-pen cultures could be maintained. Such a project might be useful in those Puget Sound embayments where surface stratification limits primary production, such as in Dabob Bay. First, however, the problems of conflicting shoreline uses, impact on the indigenous food chain, and pumping technology and costs would have to be solved.

The economic prospects of smaller scale, completely enclosed marine farms may be bright, especially when used to treat sewage and pulp effluent before it is discharged to the environment. Considerable research, in fact, has gone into the possibility of turning sewage into food via phytoplankton and shellfish in artificial ponds. There are problems with ingestion of sewage pollutants, bacteria, and viruses by the shellfish, and it remains to be demonstrated whether the products— clean water and food—will repay the cost of pumps, land, labor, and fuel, particularly in well-watered and productive regions like Puget Sound. Nevertheless, such self-contained ecosystems are ranked among the brightest of aquaculture prospects nationwide.

There may also be prospects for farming phytoplankton for direct human consumption. Natives in Africa have for hundreds of years eaten filamentous microalgae from Lake Chad. Formal research on cultivation of plankton as human food began in the 1950s, and at one point algal cultures were proposed as a food and oxygen source for space travelers. Research efforts today are leading to enclosed farming of such freshwater organisms as the cyanobacterium *Spirulina* and the green algae *Chlorella* and *Scenedesmus*, which are filtered, dried, and eaten as a food supplement in powdered form. These genera grow rapidly under harsh conditions, and can have a higher protein content and a higher per acre yield than such terrestrial crops as soybeans, but require bright sunlight and must also repay the additional costs of containers, pumps, and processing.

Algal cultures appear to be economical in arid climates where water itself is precious. They may be useful for sewage treatment in Israel, and for supplementing protein intake of both people and animals in underdeveloped countries, such as Peru, Thailand, and India. There seems little prospect for their application to the Puget Sound region. Perhaps their presence here will remain limited to the newly-developing market for *Spirulina* as a health-food supplement. Popular in Japan, the alga is cultured in Mexico, and in 1981 sold in Seattle for $30 per pound.

From all of this mixture of encouraging and discouraging news, what then is the consensus about the potential for plankton in increasing the harvest from Puget Sound and the rest of the world's oceans? The greatest potential for enhancing the harvest of food from Puget Sound seems to be in controlling the high reproductive mortality of already popular species of planktivorous animals, particularly salmon and shellfish. In addition, for the sedentary and highly efficient shellfish, there seems to be bright prospect in the creation of additional habitat for adult growth if political problems can be overcome. But these resources, even if developed to the fullest extent, will likely boost only slightly the supply of what will always be luxury foods. The great popular hope for feeding the world's starving masses remains, if not totally spurious, at least far beyond our present technological, economic, or political capabilities.

Civilization is altering its relationship with the sea and its organisms in the same way that it has changed the face of the land. We have domesticated wild animals from the dog to the chicken, and plants from the geranium to the Douglas fir. In so doing we have transformed the landscape from a wilderness into a patchwork of farms. Even more drastically, we have altered the course of evolution for entire species, intervening to perpetuate those breeds which were of value to us and neglecting or even exterminating others.

We stand today, perhaps, at the threshold of doing the same in the sea, this time with more scientific maturity and awareness of the possible consequences. Although it may seem impossible to ever change the face of the open ocean, human activities are already having an impact on coastal areas such as Puget Sound, and many difficult choices are ahead. Coastal areas can be maintained in their natural states, or developed for industry or food. Wild marine creatures can be allowed to exist in their natural habitats, or domesticated and made forever dependent on people. We may be trading in the ethic of the hunter for that of the farmer, and turning the marine wilderness into marine pastures.

Glossary

alga (plural **algae**) Plant having simple internal organization without fluid-transporting structures; unicellular or multicellular

anoxic Containing little or no oxygen

aphotic The deeper pelagic zone where sunlight is insufficient for plant growth

autotroph Organism (usually a plant) that manufactures its own food from raw materials and energy (usually sunlight)

bacterioplankton Bacteria suspended in the sea, or attached to other suspended matter

benthic Associated with the sea bottom

benthos Benthic plants and animals

binary fission Reproduction of a cell by duplication of its parts and separation into identical daughter cells

bioaccumulation Retention of pollutants in organisms at concentrations higher than in ambient water

bioassay Estimation of pollutant concentration by toxic effects on a standard test organism

biomagnification Higher body content of pollutants in animals at higher trophic levels

biomass Standing stock of organisms as measured by their collective weight

bloom Rapid, enormous increase in phytoplankton standing stock during favorable environmental conditions

B.O.D. (biological oxygen demand) Consumption of dissolved oxygen in water after addition of chemicals or organic matter

calorie Amount of energy required to heat one gram of pure water one degree Celsius

carnivore Animal that eats only other animals

Celsius (or **Centigrade**) Temperature scale having 0° at freezing point and 100° at boiling point of pure water

centimeter One-hundredth of a meter; 0.4 inches (**square centimeter** equals 0.16 square inch, **cubic centimeter** equals 0.06 cubic inch)

CEPEX (Controlled Ecosystem Pollution Experiment) Plankton growth experiment series performed in large cylindrical plastic bags deployed in Saanich Inlet, Vancouver Island, during the mid-1970s

chelation Binding of metallic ions to a complex organic molecule, affecting their solubility and activity in seawater

chelator Organic molecule causing chelation

chlorophyll *a* Principal green plant pigment, used as a measure of phytoplankton biomass

cilia Short bristly hairs that move in unison in cellular feeding or propulsion

ciliate Protozoan having distinct ciliary structure

Coelenterate Phylum of animals with stinging cells and an ancestral two-stage sedentary (hydroid) and mobile (medusoid) life cycle

colony Aggregation of organisms that could survive individually

Competitive Exclusion Principle Theory that in a constant homogenous environment, a single superior species should drive all competitors to extinction

continental shelf Shallow sea bottom fringing the continents to a depth of roughly 200 meters

copepod Small torpedo-shaped crustacean with antennae, probably the dominant metazoan in the sea
copepodite Later juvenile stage of a copepod
cyst Hard pellet formed by many neritic plants and some animals for overwintering or other long-term refuge
cytoplasm Mixture of fluid and structures in the interior of a cell exclusive of the nucleus
detritus Nonliving particles, suspended or on the bottom
diapause Period of inactivity in winter among crustaceans, analogous to hibernation in vertebrates
dinoflagellates Unicellular plants and animals possessing two characteristic flagella, and frequently outer cellulose plates
direct uptake Uptake of pollutants from water directly into organisms
El Niño Short for *El Niño de Navidad* (Child of Christmas), Peruvian term for occasional catastrophic mortalities of fish and birds, usually during December, caused by decreased upwelling
embryology Study of development of the egg from conception to mature organism
estuary Area in which fresh water meets salt water, usually in an inlet
euphotic Upper hundred meters or less of the pelagic zone, where sunlight is sufficient for plant growth
eutrophication Overenrichment of a water body by nutrients, causing nuisance phytoplankton overgrowth
fjord A long, narrow, steep-sided marine inlet, carved by a glacier, usually with a sill at its mouth
flagellum (plural **flagella**) Long whiplike cellular hair used for propulsion
fluorometer Instrument that measures chemical concentrations from how they re-emit incident fluorescent light
flushing Turnover of water in a basin by the outflow and inflow of runoff and tides
flushing time Same as **residence time**
food chain Simple model of a community having organisms assigned in a linear sequence of numbered trophic levels
food web Complex community model in which interlocking feeding patterns create a multidimensional trophic mesh without discrete levels
Foraminifera (forams) Class of protozoa having a spiral outer shell of calcium carbonate
frustule The two halves of the silica outer shell of diatoms
gram 0.035 ounce
herbivore Animal that eats only plants
hermaphrodite Animal with both sets of sexual organs
heterotroph Organism dependent for food on other organisms or their organic products
holoplankton Organisms that spend their entire lives as plankton
holotrich Group of ciliate protozoa
hydra (hydroid) The sedentary phase of the coelenterate life cycle
ichthyoplankton Planktonic fish larvae
intertidal Benthic zone between the highest and lowest tide levels
ion Chemically reactive atom or molecule having an electrical charge
kilocalorie Amount of energy required to heat one kilogram of pure water one degree Celsius; one thousand calories
kilogram One thousand grams; 2.2 pounds

kilometer One thousand meters; 0.62 miles. (**square kilometer** equals 0.4 square miles or 247 acres)

K-selected Evolved to maintain a constant population despite environmental fluctuations

Langley Radiation delivering an energy flux of one calorie per square centimeter

Langmuir cells Small-scale current spirals visible as slicks and bands paralleling the wind

larva (plural **larvae**) Earliest stage of maturity, after hatching from the egg

Limiting Nutrient Concept Theory derived from agriculture that a single nutrient, in scarcest supply relative to requirements, limits yield of a plant crop

lipid Fat or oil molecule

liter One thousand cubic centimeters; 1.06 quarts

littoral Shallowest benthic zone, often used synonymously with "intertidal"

lorica Outer sheath of a tintinnid ciliate, frequently adorned with cemented sediment

medusa (plural **medusae**) Free-floating jellyfish stage of the coelenterate life cycle, named for the resemblance of its tentacles to the snake-haired woman of mythology

meroplankton Planktonic (usually larval) stages of animals that are nekton or benthos the rest of their life cycle

meso-zooplankton Intermediate-sized zooplankton (about 0.25 to 10 millimeters) reliably sampled with a net

metazoa multicellular animals

meter 3.28 feet; 1.09 yards (**square meter** equals 10.7 square feet; **cubic meter** equals 35.3 cubic feet, 1.3 cubic yards)

metric ton One thousand kilograms: 2,200 pounds; 1.1 English tons

microalga Alga having only one or a few cells per organism

microgram One millionth of a gram

micrometer (micron) One millionth of a meter, one thousandth of a millimeter

micronekton Larger zooplankton (more than about one centimeter) with significant net-avoiding swimming ability

micro-zooplankton Smaller zooplankton (less than about 250 micrometers) poorly sampled by nets

milligram One-thousandth of a gram

millimeter One-thousandth of a meter, one-tenth of a centimeter

mitosis Duplication of a cell's genetic material prior to binary fission

morphology Internal and external structures of organisms, and their functions

nanoplankton Organisms too small (less than 20 micrometers) to be sampled even by a fine-meshed phytoplankton net

nauplius (plural **nauplii**) Earliest larval stages of copepods and many other crustaceans

neap tides Biweekly periods of weaker tidal currents and narrower tidal ranges associated with the quarter moons

nekton Pelagic animals swimming strongly enough to oppose ocean currents

neritic Shallow pelagic waters overlying the continental shelf

net plankton All plankton large enough to be caught with a fine-meshed phytoplankton net (above about 20 micrometers)

neuston Organisms associated with the very thin water-air interface

nutrient Dissolved chemical essential for plant growth

oceanic Deep pelagic waters beyond the edge of the continental shelf

oligotrich Ciliate protozoan with weak or absent sheath

omnivore Animal that eats both plants and animals

organic Pertaining to or derived from organisms; a chemical containing a carbon-hydrogen complex

organism A living entity: plant, animal, or otherwise

Paradox of the Plankton Apparent violation of the Competitive Exclusion Principle by the observed diversity of coexisting plankton species

pelagic Contained within a body of water, off the bottom

periphyton Organisms living on rooted aquatic plants

phylum (plural **phyla**) The coarsest division (after kingdom) into which organisms are categorized by their morphological and evolutionary similarities

physiology Study of metabolic systems and processes in organisms

phytoflagellates Small flagellate microalgae from several phyla, including chlorophytes, chrysophytes, and cryptophytes

phytoplankton Planktonic microalgae

plankton Aquatic organisms living unattached to the bottom and having swimming powers insufficient to resist water currents

population Standing stock of organisms as measured by their collective numbers

ppb (parts per billion) One part in 10^9 by weight, or one milligram per metric ton

ppm (parts per million) One part in 10^6 by weight, or one gram per metric ton, one milligram per kilogram

ppt (parts per trillion) One part in 10^{12}, or one microgram per metric ton

predator A carnivore, especially a mobile one

prey An animal eaten by a carnivore

primary productivity Productivity by autotrophs that supports the remainder of the community

productivity Rate of generation of new living matter by organisms

protozoa Unicellular animals

PSP (paralytic shellfish poisoning) Nerve poisoning due to eating shellfish containing toxin ingested from the dinoflagellate *Gonyaulax*

radiolarians Class of protozoa with a star-shaped silica outer skeleton

red tide Discoloration of surface water by a dense plankton bloom, including but not limited to those associated with toxicity and luminescence

residence time Average time spent by a water molecule in a basin before being flushed to sea; estimated as ratio of basin volume to inflow or outflow

r-selected Evolved for rapid population changes in response to environmental fluctuations

saxitoxin The principal neurotoxin generated by *Gonyaulax catenella* that causes paralytic shellfish poisoning

seta (plural **setae**) Small immobile bristle; its branches are "setules"

sill Shallow submerged pile of debris left across a basin by a retreating glacier; called a moraine on land

spectrophotometer Instrument that measures chemical concentrations from their absorption of known colors of light

spores Small, thinly covered bodies shed by plants for vegetative asexual reproduction or short-term refuge

spring tides Biweekly periods of stronger tidal currents and wider tidal ranges associated with the full and new moons

standing stock Quantity of organisms at a given time and place, expressed per unit area or volume

synergism Interaction of controlling factors (e.g., pollutant concentrations) in which the effect of one is enhanced by the presence of the other

taxonomy Categorizing organisms by their structural and evolutionary similarities

tintinnid Ciliate protozoan with a sturdy vase-shaped outer sheath (lorica) with cemented sediment grains

ton see **metric ton**

trophic Pertaining to the feeding relationships among organisms

trophic energy Energy fixed during primary production and available to animals as organic matter

trophic level Theoretical numerical ranking that expresses how many organisms trophic energy must pass through from its source to reach a given organism

trophic uptake Uptake of a pollutant via food

unicellular Having only one cell per organism

upwelling Upward water motion that stimulates phytoplankton growth by nutrient supply, caused by winds and currents

vacuole Hollow space in a cell for storage or buoyancy

vortex (plural **vortices**) Spiral motion

zooplankton Planktonic animals, including meroplankton

zygote Original cell of any organism, a fertilized egg

Guide to Pronunciation

Acartia   uh-CAR-sha
Aequorea   eve-QUOR-ee-uh
Amoebophyra ceratii   uh-mee-boy-FY-ruh sir-AY-she-eye
aphotic   ay-FO-tick
Brachionus   brack-ee-OH-nuss
Burien   BYOO-ree-un
Calanus   KAL-un-us
CEPEX   SEE-pecks
Ceratium   sir-AY-shum
Chaetoceros   kee-TAH-sir-us
Chaetognath   KEE-tog-nath
chelation   kee-LAY-shun
Chlorella   klor-ELL-uh
chordates   KORE-dates
cilia   SILL-ee-uh
Clione   kly-OH-nee
Coelenterates   suh-LEN-ter-ates
copepod   KOH-puh-pods
Corycaeus   koh-RIH-see-us
Coscinodiscus   kah-sin-oh-DISS-kuss
Ctenophore   TEEN-oh-for
detritus   dee-TRITE-us
diatoms   DYE-uh-toms
dinoflagellates   dye-noh-FLAJ-uh-lates
El Niño   El-NEEN-yoh
Euchaeta   you-KEET-uh
Euphausiid   you-FOW-zid
euphotic   you-FOH-tick
fjord   fyord
geoduck   GOO-ee-duck
Gonyaulax   gonn-ee-AWL-ax
halocline   HAY-loh-kline
ichthyoplankton   ICK-thee-oh-plank-tunn
larvaceans   lar-VAY-shuns
medusae   muh-DOO-see
Mytilus   MIH-tih-luss
Navicula   nuh-VICK-you-luh
Neocalanus   nee-oh-KAL-uh-nuss
Nereis   NEAR-ee-us
neritic   nuh-RIT-ick
neuston   NEW-stun
Nitzschia   NITCH-ee-uh
Noctiluca   nock-tih-LOO-kuh
Oikopleura   oy-koh-PLOO-ruh
Olisthodiscus   oh-liss-thoh-DISS-kuss
Pasiphaea   pass-ih-FEE-uh
phytoflagellates   fy-toh-FLAJ-uh-lates
Pleurobrachia   ploo-roh-BREAK-ee-uh

polychaetes   PAHL-ee-keets
Polyorchis   pahl-ee-OAR-kiss
Pseudocalanus   soo-doh-KAL-uh-nuss
pycnocline   PICK-noh-kline
Puyallup   pyoo-AL-up
Pyramimonas   puh-ram-ih-MOAN-us
Saanich   SAN-itch
Sequim   skwim
Skeletonema   skeleton-EE-muh
Spirulina   spy-roo-LEAN-uh
Synchaeta   sin-KEET-uh
Thalassiosira   thuh-lass-ee-oh-SIGH-ruh
Tintinnids   tin-TIN-idz
trophic   TROH-fick

BIBLIOGRAPHIC NOTES

The extensive collection of references consulted in preparing this book has been compiled and will be available as a Washington Sea Grant Technical Report (The Fertile Fjord: Annotated Bibliography) in late 1983. Much of the same literature has also been reviewed and discussed in Dexter et al., 1981 (see Chapter 6). Below are the sources of quotes and illustrations, along with selected references to further interesting reading.

Chapter 1   Plankton Primer

Hardy, A.C. 1965. *The Open Sea: Its Natural History.* Boston: Houghton Mifflin. Vol. 1, 335 pp.

Porter, K.G. and E.I. Robbins. 1981. Zooplankton fecal pellets link fossil fuel and phosphate deposits. *Science* 212:931–933.

Rand McNally. 1977. *The Rand McNally Atlas of the Oceans.* Chicago: Rand McNally. 208 pp.

Ryther, J.H. 1969. Photosynthesis and fish production in the sea. *Science* 166:72–76.

__________. 1970. Is the world's oxygen supply threatened? *Nature* 227:374–375.

Steinbeck, J. 1951. *The Log from the Sea of Cortez.* New York: Viking Press. 282 pp.

Tunnicliffe, V. 1981. High species diversity and abundance of the epibenthic community in an oxygen-deficient basin. *Nature* 294:354–356.

Whittaker, R.H. 1975. *Communities and Ecosystems.* Second Edition. New York: MacMillan. 385 pp.

Chapter 2   Studying Plankton

Agassiz, A. 1865. North American Acalephae. *Mem. Mus. Compar. Zool. Harvard College* 1:1–234.

Barham, E.G. 1979. Giant larvacean houses: Observations from deep submersibles. *Science* 205:1129–1131.

Cousteau, J.-Y., with J. Dugan. 1963. *The Living Sea.* New York: Harper and Row. 239 pp.

Harbison, G.R. and L.P. Madin. 1979. Diving— a new view of plankton biology. *Oceanus* 22:18–27.

Orr, M.H. 1981. Remote acoustic detection of zooplankton response to fluid processes, oceanographic instrumentation, and predators. *Can. J. Fish. Aqua. Sci.* 38:1096–1105.

Thompson, T.G. and L.D. Phifer. 1936. The plankton and the properties of the surface waters of the Puget Sound region. *Univ. Wash. Publ. Ocean.* 1:111–134.

Chapter 3   Plankton Hall of Fame

Arai, M.N. and A. Brinckmann-Voss. 1980. Hydromedusae of British Columbia and Puget Sound. *Can. Bull. Fish. Aqua. Sci.* 204. 192 pp.

Cupp, E.E. 1943. Marine plankton diatoms of the West Coast of North America. *Scripps Inst. Ocean. Bull.* 5:1–237.

Gardner, G.A. and I. Szabo. 1982. British Columbia pelagic marine Copepoda: Identification Manual and annotated bibliography. *Can. Spec. Publ. Fish. Aqua. Sci.* 62. 536 pp.

Gran, H.H. and E.C. Angst. 1931. Planktonic diatoms of Puget Sound. *Publ. Puget Sound Mar. Biol. Sta.* 7:417–456.

Smith, D.L. 1977. *A Guide to Marine Coastal Plankton and Marine Invertebrate Larvae.* Dubuque, Iowa: Kendall-Hunt Publishing Company. 161 pp.

Vinyard, W.C. 1975. *A Key to the Genera of Marine Planktonic Diatoms of the Pacific Coast of North America.* Eureka, Calif.: Mad River Press. 27 pp.

Chapter 4   Seascapes

Bienfang, P.K., P.J. Harrison, and L.M. Quarmby. 1982. Sinking rate response to depletion of nitrate, phosphate, and silicate in four marine diatoms. *Mar. Biol.* 67:295–302.

Evans, G.T. and F.J.R. Taylor. 1980. Phytoplankton accumulation in Langmuir cells. *Limnol. Oceanogr.* 25:840–845.

Silver, M.W., A.D. Shanks, and J.D. Trent. 1978. Marine snow: Microplankton habitat and source of small-scale patchiness in pelagic populations. *Science* 201:371–373.

Strickler, J.R. 1982. Calanoid copepods, feeding currents, and the role of gravity. *Science* 218:158–160.

Chapter 5   The Green Machine

Barlow, J. 1958. Spring changes in phytoplankton abundance in a deep estuary, Hood Canal, Washington. *J. Mar. Res.* 17:53–67.

Booth, B.C. 1969. Species differences between two consecutive phytoplankton blooms in Puget Sound during May, 1967. Unpubl. M.S. Thesis, Univ. Wash., Seattle. 28 pp.

Campbell, S.A., W.K. Peterson, and J.R. Postel. 1977. Phytoplankton production and standing stock in the Main Basin of Puget Sound. Final Report to the Municipality of Metropolitan Seattle, Puget Sound Interim Studies. 132 pp.

Christensen, J.P. and T.T. Packard. 1976. Oxygen utilization and plankton metabolism in a Washington fjord. *Estuar. Coast. Mar. Sci.* 4:339–347.

Coomes, C.A., C.C. Ebbesmeyer, J.M. Cox, J.M. Helseth, L.R. Hinchey, G.A. Cannon, and C.A. Barnes. (in press) Synthesis of current measurements in Puget Sound Washington. Volume 2. Indices of mass and energy inputs into Puget Sound: Runoff, air temperature, wind, and sea level. NOAA Tech. Memo.

Ebbesmeyer, C.C. and C.A. Barnes. 1980. Control of a fjord basin's dynamics by tidal mixing in embracing sill zones. *Estuar. Coast. Mar. Sci.* 11:311–330.

Ebbesmeyer, C.C. and J.M. Helseth. 1977. An analysis of primary production observed during 1966–1975 in central Puget Sound, Washington. Final Report to the Municipality of Metropolitan Seattle, Puget Sound Interim Studies. 68 pp.

English, T.S. 1979. Biological systems acoustical assessments in Port Gardner and adjacent waters. Contract report to Wash. Dept. Ecol., Olympia. 137 pp.

Feely, R.A. and M.F. Lamb. 1979. A study of the dispersal of suspended sediment from the Fraser and Skagit rivers into northern Puget Sound using LANDSAT imagery. Interagency Energy/Environment Research and Development Program Report EPA-600/7-79-165. 46 pp.

Frisch, A.S., J. Holbrook, and A.B. Ages. 1981. Observations of a summertime reversal in circulation in the Strait of Juan de Fuca. *J. Geophys. Res.* 86:2044–2048.

Harris, D.L., 1981. Tides and tidal datums in the United States. U.S. Army Corps of Engineers Coastal Engineering Research Center Special Report   7. Fort Belvoir, VA.

Harrison, P.J., J.D. Fulton, F.J.R. Taylor, and T.R. Parsons. 1983. A review of the biological oceanography of the Strait of Georgia: Pelagic environment. *Can. J. Fish. Aqua. Sci.* 40:1064-1094.

Munson, R.E. 1970. The horizontal distribution of phytoplankton in a bloom in Puget Sound during May, 1969. Unpubl. M.S. Thesis, Univ. Wash., Seattle. 13 pp.

Parsons, T.R., J. Stronach, G.A. Borstad, G. Louttit, and R.I. Perry. 1981. Biological fronts in the Strait of Georgia, British Columbia, and their relation to recent measurements of primary productivity. *Mar. Ecol. Prog. Ser.* 6:237–242.

Parsons, T.R., R.I. Perry, E.D. Nutbrown, W. Hsieh, and C.M. Lalli. 1983. Frontal zone analysis at the mouth of Saanich Inlet, British Columbia, Canada. *Marine Biology* 73:1–5.

Stockner, J.G., D.D. Cliff, and K.R.S. Shortreed. 1979. Phytoplankton ecology of the Strait of Georgia, British Columbia. *J. Fish. Res. Bd. Can.* 36:657–666.

Takahashi, M., J.E. Barwell-Clarke, F. Whitney, and P.A. Koeller. 1978. Winter condition of marine plankton populations in Saanich Inlet, B.C., Canada. I. Phytoplankton and its surrounding environment. *J. Exp. Mar. Biol. Ecol.* 31:283–301.

Thomson, R.E. 1981. Oceanography of the British Columbia Coast. *Can. Spec. Publ. Fish. Aqua. Sci.* 56. 291 pp.

U.S. Environmental Data Service. Climatological Data for Washington. National Weather Service, Seattle.

Winter, D., K. Banse, and G. Anderson. 1975. The dynamics of phytoplankton blooms in Puget Sound, a fjord in the northwestern United States. *Mar. Biol.* 29:139–176.

Chapter 6 The Fish Factory

Banse, K. 1982. Mass-scaled rates of respiration and intrinsic growth in very small invertebrates. *Mar. Ecol. Progr. Ser.* 9:281–297.

Chester, A.J., D.M. Damkaer, D.B. Dey, G.A. Heron, and J.D. Larrance. 1980. Plankton of the Strait of Juan de Fuca, 1976–1977. Interagency Energy/Environment Research and Development Program Report EPA-600/7-80-032. 64 pp.

Cooney, R.T. 1971. Zooplankton and micronekton associated with a diffuse sound-scattering layer in Puget Sound, Washington. Ph.D. Thesis, Univ. Wash., Seattle. 208 pp.

Copping, A.E. and C.J. Lorenzen. 1980. Carbon budget of a marine phytoplankton-herbivore system with a carbon-14 as a tracer. *Limnol. Oceanogr.* 25:873–882.

Corkett, C.J. and I.A. McLaren. 1978. The biology of *Pseudocalanus*. *Adv. Mar. Biol.* 15:1–231.

Dexter, R.N., D.E. Anderson, E.A. Quinlan, L.S. Goldstein, R.M. Strickland, S.P. Pavlou, J.R. Clayton, Jr., R.M. Kocan, and M. Landolt. 1981. A summary of knowledge of Puget Sound related to chemical contaminants. NOAA Tech. Memo. OMPA-13. 435 pp.

English, T.S. and R.E. Thorne. 1977. Acoustic and net surveys of fishes and zooplankton. Final Report to the Municipality of Metropolitan Seattle, Puget Sound Interim Studies. 64 pp.

Greve, W. and T.R. Parsons. 1977. Photosynthesis and fish production: Hypothetical effects of climatic change and pollution. *Helg. Wissenschaft. Meeresunt.* 30:666–672.

Grice, G.D., R.P. Harris, M.R. Reeve, J.F. Heinbokel, and C.O. Davis. 1980.

Large-scale enclosed water-column ecosystems. An overview of Foodweb I, the final CEPEX experiment. *J. Mar. Biol. Assoc. U.K.* 60:401–414.

Grice, G.D. and M.R. Reeve, Eds. 1982. *Marine Mesocosms.* New York: Springer-Verlag. 430 pp.

Hebard, J.F. 1956. The seasonal variation of zooplankton in Puget Sound. M.S. Thesis, Univ. Wash., Seattle. 64 pp.

Huntley, M.E. and L.A. Hobson. 1978. Medusa predation and plankton dynamics in a temperate fjord, British Columbia. *J. Fish. Res. Bd. Can.* 35:257–261.

King, K.R. 1979. The life history and vertical distribution of the chaetognath, *Sagitta elegans*, in Dabob Bay, Washington. *J. Plankton Res.* 1:153–167.

__________. 1981. The quantitative natural history of *Oikopleura dioica* (Urochordata: Larvacea) in the laboratory and in enclosed water columns. Ph.D. Thesis, Univ. Wash., Seattle. 152 pp.

Koeller, P.A., J.E. Barwell-Clarke, F. Whitney, and M. Takahashi. 1979. Winter condition of marine plankton populations in Saanich Inlet, B.C., Canada. III. Mesozooplankton. *J. Exp. Mar. Biol. Ecol.* 37:161–174.

Landry, M.R. 1981. Switching between herbivory and carnivory by the planktonic marine copepod *Calanus pacificus. Mar. Biol.* 65:77–82.

Mackas, D.L., G.C. Louttit, and M.J. Austin. 1980. Spatial distribution of zooplankton and phytoplankton in British Columbia coastal waters. *Can. J. Fish. Aqua. Sci.* 37:1476–1487.

Mauchline, J. 1980. The biology of mysids and euphausiids. *Adv. Mar. Biol.* 18:1–682.

Maynard, S. 1972. A study of the stomach contents of five fish species collected in Carr Inlet, Puget Sound. Unpubl. M.S. Thesis, Univ. Wash. 96 pp.

Ross, R.M., K.L. Daly, and T.S. English. 1982. Reproductive cycle and fecundity of *Euphausia pacifica* in Puget Sound, Washington. *Limnol. Oceanogr.* 27:304–314.

Runge, J.A. 1981. Egg production of *Calanus pacificus* Brodsky and its relationship to seasonal changes in phytoplankton availability. Ph.D. Thesis, Univ. Wash., Seattle. 116 pp.

Shuman, F. 1978. The fate of phytoplankton chlorophyll in the euphotic zone—Washington coastal waters. Ph.D. Thesis, Univ. Wash., Seattle. 243 pp.

Simenstad, C.A., B.S. Miller, C.F. Nyblade, K. Thornburgh, and L.J. Bledsoe. 1979. Food web relationships of northern Puget Sound and the Strait of Juan de Fuca. Interagency Energy/Environment Research and Development Program Report EPA 600/7-79-259. 335 pp.

Steele, J.H. 1974. *The Structure of Marine Ecosystems.* Cambridge, Mass.: Harvard University Press. 128 pp.

Takahashi M. and K.D. Hoskins. 1978. Winter conditions of marine plankton populations in Saanich Inlet, British Columbia, Canada. II. Micro-zooplankton. *J. Exp. Mar. Biol. Ecol.* 32:27–37.

Welschmeyer, N.A. 1982. The dynamics of phytoplankton pigments: Implications for zooplankton grazing and phytoplankton growth. Ph.D. Thesis, Univ. Wash., Seattle. 176 pp.

Yen, J. 1982. Predatory feeding ecology of *Euchaeta elongata* Esterly, a marine planktonic copepod. Ph.D. Thesis, Univ. Wash., Seattle. 130 pp.

Chapter 7 Plankton and Pollution

American Petroleum Institute. 1979. *1979 Oil Spill Conference.* Washington, D.C.: API. 728 pp.

Barrick, R.C. 1982. Flux of aliphatic and polycyclic aromatic hydrocarbons to central Puget Sound from Seattle (Westpoint) primary sewage effluent. *Environ. Sci. Tech.* 16:682–692.

Beak Consultants. 1975. Oil pollution and the significant biological resources of Puget Sound. Final Report to Wash. Dept. Ecol., Olympia. 3 volumes.

Brown, L.R. 1980. Fate and effect of oil in the aquatic environment—Gulf Coast region. EPA Rep. 600/3-80-058a. 101 pp.

Brown, D.A. and T.R. Parsons. 1978. Relationships between cytoplasmic distribution of mercury and toxic effects to zooplankton and chum salmon (*Oncorhynchus keta*) exposed to mercury in a controlled ecosystem. *J. Fish. Res. Bd. Can.* 35:880–884.

Cardwell, R.D., C.E. Woelke, M.I. Carr, and E.W. Sanborn. 1979. Toxic substance and water quality effects on larval marine organisms. Wash. Dept. Fish. Tech. Rep. 45, 71 pp.

Carpenter, R., M.L. Peterson, and R.A. Jahnke. 1978. Sources, sinks, and cycling of arsenic in the Puget Sound region. *In* M.L. Wiley, Ed., *Estuarine Interactions*. New York: Academic Press. Pp. 459–480.

Corner, E.D.S. 1978. Pollution studies with marine plankton. Part I. Petroleum hydrocarbons and related compounds. *Adv. Mar. Biol.* 15:289–380.

Davies, A.G. 1978. Pollution studies with marine plankton. Part 2. Heavy Metals. *Adv. Mar. Biol.* 15:381–508.

Edmonson, W.T. and J.T. Lehman. 1981. The effect of changes in the nutrient income on the condition of Lake Washington. *Limnol. Oceanogr.* 26:1–29.

GESAMP (Joint Group of Experts on the Scientific Aspects of Marine Pollution). 1977. Impact of Oil on the Marine Environment. Reports and Studies #6, Food and Agriculture Organization of the United Nations, 250 pp.

Geyer, R.A., Ed. 1980. *Marine Environmental Pollution. 1. Hydrocarbons*. Amsterdam: Elsevier Scientific Publishing Co. 591 pp.

Goldberg, E.D., V.T. Bowen, J.W. Farrington, G. Harvey, J.H. Martin, P.L. Parker, R.W. Risebrough, W. Robertson, E. Schneider, and E. Gamble. 1978. The mussel watch. *Environ. Conserv.* 5:101–125.

Grice, G.D. and D.W. Menzel. 1977. Controlled ecosystem pollution experiment: effect of mercury on enclosed water columns. VIII. Summary of results. *Mar. Sci. Comm.* 4:23–31.

Jernelov, A. and O. Linden. 1981. IXTOC I: A case study of the world's largest oil spill. *Ambio* 10:299–306.

Kineman, J.J., R. Elmgren, and S. Hansson, Eds. 1980. The *Tsesis* Oil Spill. NOAA Office of Marine Pollution Assessment, Boulder, CO.

Krebs, C.T. and K.A. Burns. 1977. Long-term effects of an oil spill on populations of the salt-marsh crab *Uca pugnax. Science* 197:484–487.

Lannergren, C. 1978. Net- and nanoplankton: Effects of an oil spill in the North Sea. *Botanica Marina* 21:353–356.

Mackie, P.R., R. Hardy, and K.J. Whittle. 1978. Preliminary assessment of the presence of oil in the ecosystem at Ekofisk after the blowout, April 22–30, 1977. *J. Fish. Res. Bd. Can.* 35:544–551.

Malins, D.C., Ed. 1977. *Effects of Petroleum on Arctic and Subarctic Marine Environments and Organisms*. Volume 1: Nature and Fate of Petroleum. 321 pp. Volume 2: Biological Effects. 500 pp. New York: Academic Press.

Malins, D.C., B.B. McCain, D.W. Brown, A.K. Sparks, H.O. Hodgins, and S.-L. Chan. 1982. Chemical Contaminants and Abnormalities in Fish and Invertebrates from Puget Sound. NOAA Tech. Memo. OMPA-19. 168 pp.

Menzel, D.W. 1977. Summary of experimental results: Controlled ecosystem pollution experiment. *Bull. Mar. Sci.* 27:142–145.

Middleditch, B.S., Ed. 1981. *Environmental Effects of Offshore Oil Production.*
New York: Plenum Press. 446 pp.

O'Connors, H.B., Jr., C.F. Wurster, C.D. Powers, D.C. Biggs, and R.G. Rowland.
1978. Polychlorinated biphenyls may alter marine trophic pathways by reducing photoplankton size and production. *Science* 201:737–739.

Officer, C.B. and J.H. Ryther. 1977. Secondary sewage treatment versus ocean
outfalls: An assessment. *Science* 197:1056–1060.

Parsons, T.R., L.J. Albright, and J. Parslow. 1980. Is the Strait of Georgia becoming more eutrophic? *Can. J. Fish. Aqua. Sci.* 37:1043–1047.

Phillips, D.J.H. 1978. Use of biological indicator organisms to quantitate organochlorine pollutants in aquatic environments—a review. *Environ. Poll.*
16:167-229.

Rasmussen, L.F. and D.C. Williams. 1975. The occurrence and distribution of
mercury in marine organisms in Bellingham Bay. *Northwest Sci.* 49:87–94.

Shea, G.G., C.C. Ebbesmeyer, Q.J. Stober, K. Pazera, J.M. Cox, S. Hemingway,
J.M. Helseth, and L.R. Hinchey. 1981. History and Effect of Pulp Mill Effluent
Discharges, Bellingham, Washington. Report to U.S. Department of Justice
and Environmental Protection Agency by Northwest Environmental Consultants, Inc. 491 pp.

Spooner, M.F. 1978. Editorial introduction (to special issue on *Amoco Cadiz*).
Mar. Poll. Bull. 9:281–284.

Stockner, J.G., D.D. Cliff, and D.B. Buchanan. 1977. Phytoplankton production
and distribution in Howe Sound, British Columbia: A coastal marine embay-
ment-fjord under stress. *J. Fish. Res. Bd. Can.* 34:907–917.

Welch, E.B., R.M. Emery, R.I. Matsuda, and W.A. Dawson. 1972. The relation of
periphytic and planktonic algal growth in an estuary to hydrographic factors.
Limnol. Oceanogr. 17:731–737.

Chapter 8   Red Tides

Cardwell, R.D., C.E. Woelke, M.I. Carr, and E. Sanborn. 1977. Evaluation of water quality of Puget Sound and Hood Canal in 1976. NOAA Tech. Memo.
ERL/MESA-21, 36 pp.

Cardwell, R.D., S. Olson, M.I. Carr, and E.W. Sanborn. 1979. Causes of oyster
larvae mortality in South Puget Sound. NOAA Tech. Memo. ERL/MESA-39,
73 pp.

MacDonald, E.M. 1970. The occurrence of paralytic shellfish poison in various
species of shore animals along the Strait of Juan de Fuca in the state of Washington. M.S. Thesis, Univ. Wash., Seattle. 65 pp.

Norris, L. and K.K. Chew. 1975. Effect of environmental factors on growth of
Gonyaulax catenella. In R. LoCicero, Ed. *Toxic Dinoflagellate Blooms.*
Wakefield, Mass.: Massachusetts Science and Technology Foundation, pp.
143–152.

Quayle, D. 1969. Paralytic Shellfish Poisoning in British Columbia. *Fish. Res.
Brd. Can. Bull.* 168. 68 pp.

Steidinger, K.A. and K. Haddad. 1981. Biologic and hydrographic aspects of red
tides. *Bioscience* 31:814–819.

Taylor, D.L. and H.H. Seliger, Eds. 1979. *Toxic Dinoflagellate Blooms.* New
York: Elsevier/North Holland. 534 pp.

Taylor, F.J.R. 1968. Parasitism of the toxin-producing dinoflagellate *Gonyaulax
catenella* by the endoparasitic dinoflagellate *Amoebophyra ceratii. J. Fish.
Res. Bd. Can.* 25:2241–2245.

Welling, K. 1982. The effects of extracellular products of *Olisthodiscus luteus* Carter on the survival of Pacific oyster larvae (*Crassostrea gigas* Thunberg) and on the growth of larval associated bacteria. M.S. Thesis, Univ. Wash., Seattle. 66 pp.

Chapter 9   Marine Pastures
Goldman, J.C. 1979. Outdoor algal mass cultures. I. Applications. *Water Res.* 13:1–19.
Omori, M. 1978. Zooplankton fisheries of the world: A review. *Mar. Biol.* 48:199–205.
Soeder, S.J. and R. Binsack, Eds. 1978. Microalgae for food and feed. *Ergebnisse der Limnol. Archiv fur Hydrobiol. Beiheft 11.* 300 pp.

Index

Italic page numbers denote illustrations.

Other Books in this Series

The Water Link: A History of Puget Sound as a Resource
Daniel Jack Chasan

Governing Puget Sound
Robert L. Bish

Marine Birds and Mammals of Puget Sound
Tony Angell and Kenneth C. Balcomb III

The Coast of Puget Sound: Its Processes and Development
John Downing